Mo n

A321
A320
A319
A319 AIRBUS
A318
A318 AIRBUS

John Nixon

Modern English for Aeronautics and Space Technology

Textbook, Excercises, Questions, Vocabulary, Answer Key

HANSER

John D. Nixon, M. A.
Akademischer Oberrat
University of Stuttgart
Language Centre

Bibliographic information published by the Deutsche Nationalbibliothek

The Deutsche Nationalbibliothek lists this publication in the Deutsche Nationalbibliografe; detailed bibliographic data are available in the Internet at http://dnb.d-nb.de

ISBN 978-3-446-42229-2
E-Book-ISBN 978-3-446-42834-8

Cover Illustrations: AeroLogic GmbH, Engine Alliance, NASA

www.hanser.de
Project Editor: Jochen Horn
Production: Katrin Wulst
Typesetting by: le-tex publishing services GmbH, Leipzig
Cover Realization: Stephan Rönigk
Printed and bound by Druckhaus "Thomas Müntzer" GmbH, Bad Langensalza
Printed in Germany

Preface

This book started as a collection of exercises and texts that I had developed for two of my English for Specific Purposes (ESP) courses in the English Department at the University of Stuttgart's Language Centre – English for Aeronautics and English for Space Engineering. The texts that formed the foundation of my readers had been drawn from various sources, such as scientific magazines, academic journals, textbooks and the Internet, and served as authentic material, through which my students could hone their English language skills. Given the lack of teaching materials for these highly technical ESP courses, I decided to compile the material I had already developed and hence the genesis of this textbook.

This book is intended as a textbook to accompany ESP courses at post-secondary institutions across the German-speaking world, but, if one disregards the translation exercises involving German, it could easily be used for English for Aerospace classes anywhere. Additionally, this textbook could be used by engineers already working in the field of aerospace engineering as part of a structured course or a self-study module.

While there is no pre-requisite *per se* for the course in which this book is meant to be used, students are generally required to have a firm grasp of the English language. The level that has been aimed for is C1 (Common European Framework of Reference for Languages) or UNIcert Level III.

As the material has been taken from a number of sources in its original context, the variety of English might vary from text to text. Consequently, the British English "centre", for example, might figure next to its American equivalent "center". Given that I hail from Canada, where I grew up recognizing that English cannot be straitjacketed into one form, the debate over British and American English has taken a back seat during the development of this book. To that end, I have striven to provide examples of texts representative of "international English", as most native speakers of English would understand it. In some cases, where there are competing variants, generally British and American, these have been provided.

This book is a combination of reader and workbook. Space has generally been provided for students to enter their answers into the book as they work through the exercises under the guidance of an instructor or on their own as part of a self-study module.

Although there is a large focus on reading comprehension, the expansion of specialized and academic vocabulary as well writing skills, there are exercises throughout the textbook that provide practice in listening comprehension and speaking. A few task-oriented exercises have also been provided to allow the students to employ all of their English-language resources in authentic, real-life situations.

I cannot vouch for the accuracy of the information provided in the material that I have gathered, but I have endeavoured to ensure that the material has been drawn from reputable sources.

Thanks goes to a number of individuals who provided me with help and advice along the way. In particular, I would like to thank *Paula Cilinghir*, who was of enormous assistance with the layout and graphics. I am also indebted to my aerospace students, without whose enthusiasm for English and for their field of study this book would never have come about.

Lastly, I would like to thank the Carl Hanser Verlag for agreeing to publish this manuscript as well as for their ideas regarding formatting, graphics and illustrations.

Stuttgart, May 2011 John D. Nixon

Contents

1 Aerodynamic Forces

1.1 Introductory Discussion

There are numerous forces acting on objects in everyday life.

For example, which forces are at play on a boat floating on water?

What about a rock on an incline that doesn't roll down the slope?

Which of these same forces operate on an aircraft in flight? Are there any others?

Discuss the preceding introductory questions and then fill in the table below.

Object	Forces acting on it
Boat	
Rock	
Airplane	

1.2 Exercises

1.2.1 Exercise A

Read the following text and fill in the blanks with the words presented here.

denotes	resists	magnitude	adjust
opposing	overcome (x2)	payload	generate
consumes	varied		

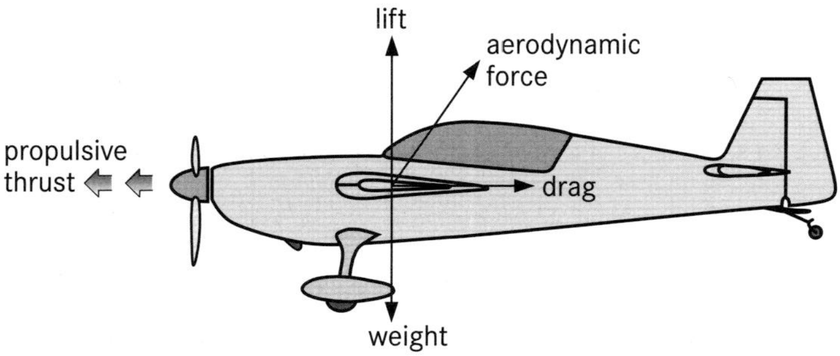

Fig. 1.1 Overview of the four aerodynamic forces acting on an airplane

Weight

Weight is a force that is always directed toward the center of the earth. The magnitude of the weight depends on the mass of all the airplane parts, plus the amount of fuel, plus any payload (1) on board (people, baggage, freight, etc.). The weight is distributed throughout an airplane. But we can often think of it as collected and acting through a single point called the center of gravity. In flight, an airplane rotates about the center of gravity.

Flying encompasses two major problems: overcoming the weight of an object by some opposing force, and controlling the object in flight. Both of these problems are related to the object's weight and the location of the center of gravity. During a flight, an airplanes' weight constantly changes as the aircraft consumes (2) fuel. The distribution of the weight and the center of gravity also changes. So, the pilot must constantly adjust (3) the controls to keep the airplane balanced or trimmed.

Lift

To overcome (4) the weight force, airplanes generate an opposing (5) force called lift. Lift is generated by the motion of the airplane through the air and is an aerodynamic force. "Aero" stands for air and "dynamic" denotes (6) motion. Lift is directed perpendicular to the flight direction. The magnitude of the lift depends on several factors including the shape, size and velocity of the aircraft. As with weight, each part of the aircraft contributes to the aircraft lift force. Most of the lift is generated by the wings. Aircraft lift acts through a single point called the center of pressure. The center of pressure is defined just like the center of gravity, but using the pressure distribution around the body instead of the weight distribution.

Drag

As an airplane moves through the air, there is another aerodynamic force present. The air resists (7) the motion of the aircraft and the resistance force is called drag. Drag is directed along and opposed to the direction of flight. As with lift, there are many factors that affect the magnitude (8) of the drag force including the shape of the aircraft, the "stickiness" of the air, and the velocity of the aircraft. All of the individual components' drags are combined into a single aircraft drag magnitude. And like lift, drag acts through the aircraft center of pressure.

Thrust

To overcome (9) drag, airplanes use a propulsion system to generate (10) a force called thrust. The direction of the thrust force depends on how the engines are attached to the aircraft. On many planes, two turbine engines are located under the wings and parallel to the body, with thrust acting along the body centreline. On some aircraft, such as the Harrier, the thrust direction can be varied (11) to help the airplane take off over a very short distance. The magnitude of the thrust depends on many factors associated with the propulsion system including the type of engine, the number of engines and the throttle setting.

(adapted from NASA Glenn Research Center, Aerodynamics Index)

1.2.2 Exercise B—Vocabulary

Look at the previous text extracts and try to answer the questions below with a partner.

1) What is the difference between weight and mass?

....

....

....

....

....

2) What is the difference between speed and velocity?

3) What is meant by perpendicular? (see text on **Lift** above)

1.2.3 Exercise C—Translation

Translate the following text into English.

Auf ein Flugzeug, das sich gerade im Flug befindet, wirken immer vier Kräfte ein: der Auftrieb, die Schwerkraft (also das Gewicht des Flugzeugs), der Schub (oder auch „Vortrieb“ genannt) und der Widerstand. Wie man sich vorstellen kann, sind Auftrieb und Schwerkraft gegeneinander gerichtet, und dasselbe gilt auch für den Vortrieb (Schub) und den Widerstand. Der Auftrieb ist das wesentlichste Element beim Fliegen. Während des Fluges wirken aber noch andere Kräfte auf das System ein und in deren Zusammenspiel ergibt sich das Flugverhalten des Geräts.

(adapted from Drachen- und Gleitschirmclub, www.dgcb.de)

1.3 Why an Aircraft Flies

1.

As an aircraft is “heavier than air”, it needs an upward force to keep it aloft. This force is provided by the “lift” developed by the supporting surfaces (wings) and is directed at right angles to the direction of movement. In addition, the air offers a certain frictional resistance, called “drag”. By suitable design of the cross-sectional shape of the wing, the drag can be kept small in relation to the lift. A shape of this kind is known as an airfoil section (Fig. 1.2).

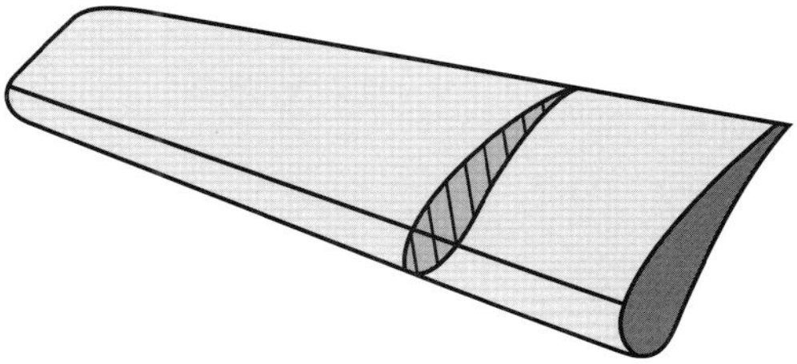

Fig. 1.2 Airfoil section

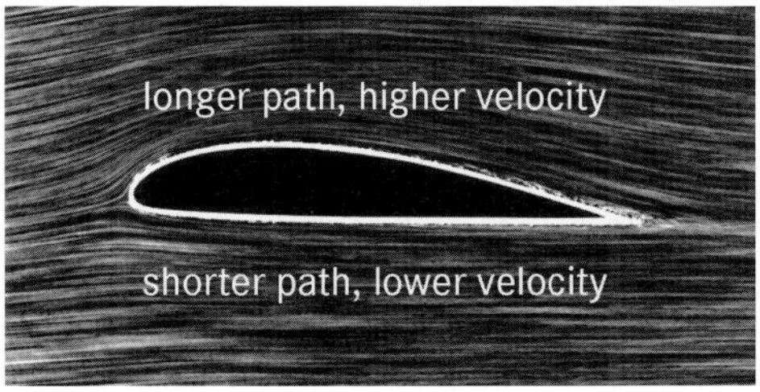

Fig. 1.3 Airflow around an airfoil

2.

When the wing of aircraft moves forward through the air, the flow of air along the lower surface arrives at the trailing edge before the flow along the upper surface. The lower surface flow attempts to expand around the trailing edge. As a result of this, a vortex is formed. The rotation of this vortex accelerates the upper surface flow, so that the length of time required for a particle of air to move from the leading edge to the trailing edge becomes the same for the upper and the lower surface flow. The increased velocity of the upper surface flow eliminates the formation of a vortex by the lower surface air at the trailing edge, and it produces a lower pressure at each point on the upper surface than exists at the corresponding points on the lower surface. It is this difference in pressure that produces the lift. The distribution of lift along the cross-section of a wing is illustrated by the pressure-distribution diagram (Fig. 1.4). The magnitude of the forces changes with the angle of attack (or angle of incidence), i. e. the angle between the direction of the air flow and the chord line of the wing (Fig. 1.5). The resultant aerodynamic force acts at the centre of pressure (Fig. 1.4); its position varies with the angle of attack. The stability of an aircraft is significantly determined by the displacement of the centre of pressure. With an increasing angle of attack, this point moves forward. When the angle is increased beyond the value that produces maximum lift, “stall” occurs: this results in loss of flying speed and lift and finally loss of control; the air flow detaches itself from the upper surface (Fig. 1.6).

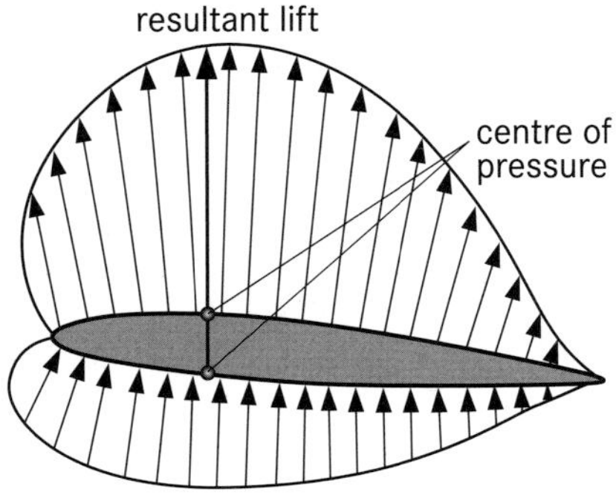

Fig. 1.4 Pressure distribution diagram

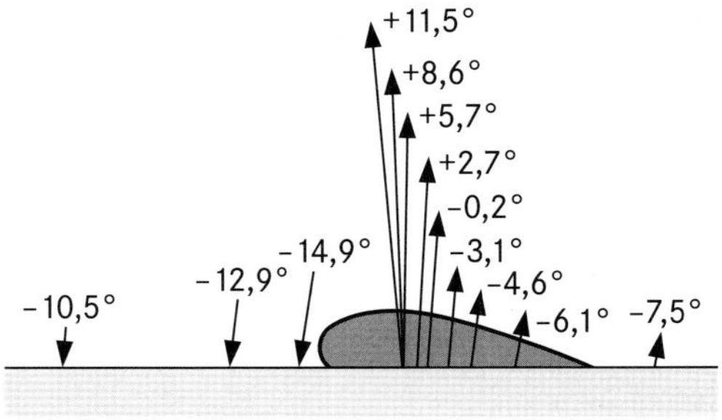

Fig. 1.5 Variation of forces and position of centre of pressure with angle of attack

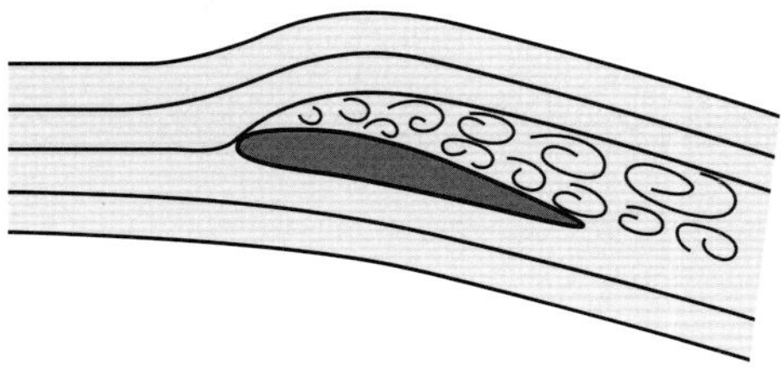

Fig. 1.6 Formation of stall

3.
The airfoil section is so shaped as to present minimum air resistance at the design speed of the aircraft and at the same time provide the necessary amount of lift. Figs. 1.7a and 1.7b represent the wing sections of a cargo-carrying aircraft and a faster aircraft respectively. The lift provided by the thick highly curved wing is about half as much again as that of the thinner and flatter wing, but its drag is about twice as high.

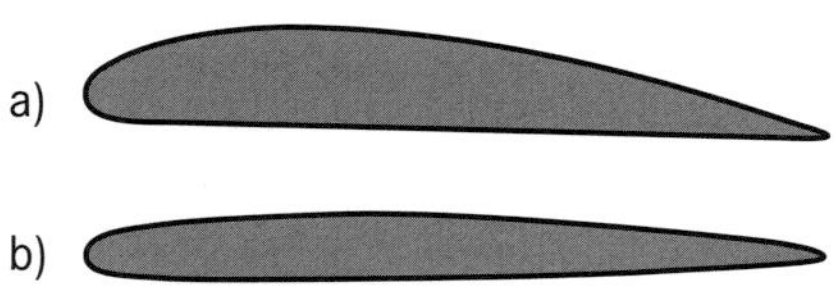

Fig. 1.7 Different airfoils

4.
The principles of airfoil design are also applicable to the propeller blades. The function of the propeller is to convert the torque developed by the engine into a propulsive thrust to drive the aircraft forward. This thrust is produced by acceleration of the air around the propeller. Since the velocity at each blade section is a function of radius, the blades are twisted to maintain a favourable angle of attack all along the blade. The principal forces and velocities associated with the action of the propeller are shown in Fig. 1.8. The pitch angle of the propeller blade corresponds to the angle of attack of the wing: it is the angle between the blade chord line and the plane of rotation. A variable-pitch propeller is designed to maintain propeller efficiency as the forward velocity changes. The pitch setting can be changed while the propeller is rotating.

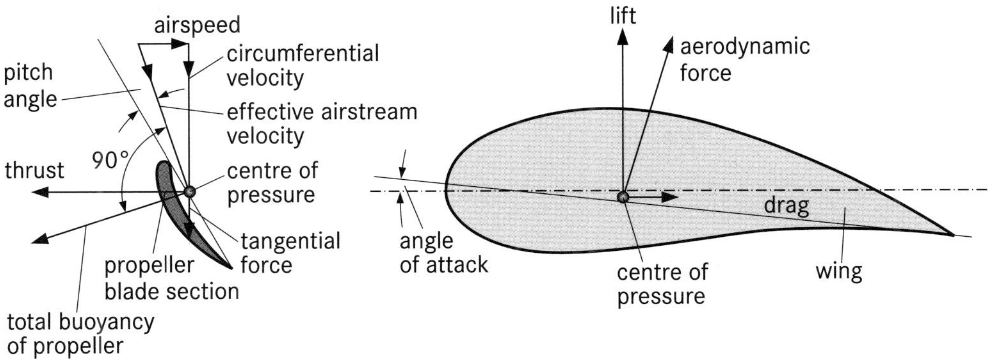

Fig. 1.8 Cross-section of an airfoil and the various forces at work

1.3.1 Exercise A: Comprehension

Using the text above, answer the following questions using full sentences.

1) Define the concepts "lift" and "drag".

....

....

....

....

....

2) How is lift produced?

....

....

....

....

....

3) What factor determines the stability of an aircraft?

....

....

....

....

4) How does the wing shape of a cargo plane differ from that of a faster aircraft?

....

....

....

....

....

5) What is the function of a propeller?

....

....

....

....

6) What is a variable-pitch propeller?

....

....

....

....

1.3.2 Exercise B: Vocabulary

Paragraph One

a) What does 'aloft' mean?

in the air

b) What German word is 'loft' related to etymologically?

Luft

c) Can you think of other words with the same type of structure 'a' + 'noun'/'adj' = 'adj' or 'adverb'

ashore, abroad, asleep, alife

Paragraph Two

d) What is the opposite of 'accelerate'? to speed up

decelerate = to slow down

e) What is the plural of 'aircraft'?

aircraft

Paragraph Three

f) What is the purpose of the word 'respectively' in the second sentence?

bzw. (Im Deutschen steht bzw. an anderer Stelle!)
anstatt, and

g) Create a sentence using the structure 'so + adj. + as to' found in this paragraph.

would you be so kind as to do this for me?

Paragraph Four

h) What does the word 'maintain' mean in the text here? What is another meaning, which is not found in the text?

to keep (here!),
to repair, to inspect

i) Define 'torque'.

j) Word formation: propeller → .... (verb)

→ .... (adj.)

1.3.3 Exercise C: Fill in the Blanks

Read this aerodynamics course description from the University of Michigan. Fill in the missing words with the words below. Note that there are not enough words and that you will have to come up with a few on your own. Refer back to the text you have just read to find the answers.

aerodynamics **separation** **calculated** **concepts**

This course deals with the fundamental concepts in aerodynamics. Students learn how airfoils produce lift, and how the pressure distribution about an airfoil can be calculated. The course also introduces the boundary-layer concept, how boundary layers lead to drag, and what makes them prone to instability or separation. It covers the effects of the wing planform (≙ airfoil when looking from above) shape on lift and drag and includes an introduction to airfoil design and high-speed aerodynamics.

to be prone to
≙ anfällig für

1.3.4 Exercise D: Writing

In your own words and using complete sentences, try to explain the function of wings and engines on an airplane. Focus on the role of these airplane parts as they relate to aerodynamic forces.

....

....

....

....

....

....

....

....

....

....

Fig. 1.9 GP 7000

1.4 How Does an Aircraft Really Fly?

1.

Lift is the force that holds an aircraft in the air. How is lift generated? There are many explanations for the generation of lift found in encyclopedias, in basic physics textbooks, and on Web sites. The proponents of the arguments usually fall into two camps: (1) those who support the „Bernoulli“ position that lift is generated by a pressure difference across the wing, and (2) those who support the „Newton“ position that lift is the reaction force on a body caused by deflecting a flow of gas.

2.

When a gas flows over an object, or when an object moves through a gas, the molecules of the gas are free to move about the object; they are not closely bound to one another as in a solid. Because the molecules move, there is a velocity associated with the gas. Within the gas, the velocity can have very different values at different places near the object. Bernoulli's equation relates the pressure in a gas to the local velocity; so as the velocity changes around the object, the pressure changes as well. Adding up (integrating) the pressure variation times the area around the entire body determines the aerodynamic force on the body. The lift is the component of the aerodynamic force which is perpendicular to the original flow direction of the gas. The drag is the component of the aerodynamic force which is parallel to the original flow direction of the gas.

Adding up the velocity variation around the object instead of the pressure variation also determines the aerodynamic force. The integrated velocity variation around the object produces a net turning of the gas flow. From Newton's Third Law of Motion, a turning action of the flow will result in a re-action (aerodynamic force) on the object.

Fig. 1.10 Airbus A350

So both „Bernoulli“ and „Newton“ are correct. Integrating the effects of either the pressure or the velocity determines the aerodynamic force on an object. We can use equations developed by each of them to determine the magnitude and direction of the aerodynamic force.

(adapted from www.grc.nasa.gov)

1.4.1 Exercise A: Diagram

Draw and label a diagram illustrating how lift is generated on an airfoil using the Newtonian explanation.

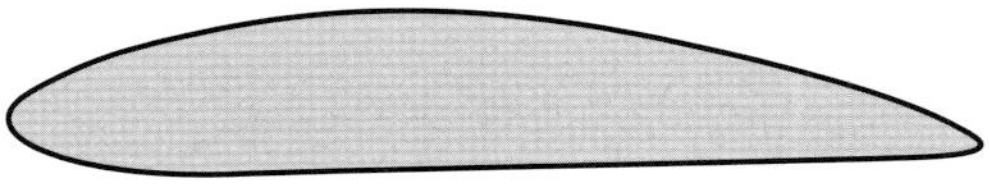

Fig. 1.11

1.4.2 Exercise B: Discussion Questions

With a partner discuss the following questions:

Is this the complete explanation of how airplanes fly?

What are the Navier-Stokes equations?

If scientists and engineers have a hard time predicting how objects will behave in a moving liquid such as air, what consequences does that have for airplane manufacturers?

1.4.3 Language Corrections

Look at the sentences below and make all the necessary corrections. The mistakes can be related to either grammar, vocabulary or spelling.

1) The wings produce lift if the aircraft moves thru the air.

2) Engines are most often directed so that thrust works in opposite direction than the velocity of flight.

3) The air on the upper side moves faster. By this the pressure becomes smaller.

4) The movement of the wing, and so the plane, has to be fast enough to produce a difference in pressure.

5) This airfoil should produce a small as possible amount of drag.

6) Lift is an upward force that overcomes the gravity.

7) Drag and thrust are opposing each other.

8) The difference in pressure generates the force mentioned on the beginning.

9) But this principle is only possible when the aircraft is moving forward.

10) On conventional aircraft designs, the ailerons and flaps are also attached to the wings.

11) The wings of an airplane produce the mean part of the lift.

12) Engines on an airplane produces the necessary thrust to overcome drag.

13) Wings accelerate the air flow on the upper part of the airfoil.

14) Drag must be opposed by propulsive thrust, as the plane otherwise would decelerate and loose height.

15) On the upper side of the wing, the air has to go a larger distance.

16) The air moves faster and this results an area of low pressure.

17) Lift results of the difference between the pressure at the upper and lower surface of an airfoil.

18) Because of the higher velocity on the upper surface you receive a lower pressure.

19) You must reach a velocity which gives the possibility to lift and to keep the aircraft aloft.

20) Engines have the function of making the plane move forward.

Fig. 1.12 Airbus A320neo

1.5 James Bond 007—Reading Numbers Aloud in English

1.5.1 Multiple Digits

a) In English, the digits of phone numbers, credit cards, bank accounts, etc. are usually read out one by one.

Hence, the credit card number **4980 7065 3245 9128** is read as

four nine eight oh seven oh six five three two four five nine one two eight

It is not customary to group numbers together as in German, e.g.

forty-nine eighty seventy sixty-five...

b) In British English, quite often identical digits side by side are read as double, triple (treble)

How would you say the following?

James Bond 007 → [double oh seven] } British

777-5672 → [triple (or treble) seven five six seven two]

1.5.2 Zero

There are many different ways of saying **zero** in English.

a) **zero and oh**

These two forms are the most common.

1) When numbers are said aloud one at a time, the British tend to say 'oh' for 0 and North Americans 'zero'.

e. g. "My account number is four one three oh six." (in the UK)

"My account number is four one three zero six." (in North America).

However, numbers may be 'chunked' together. That is, a North American may say 'highway four oh one' for 'Hwy. 401' or the telephone area code for Washington D.C. is 'two oh two'. This occurs when the speaker considers the group of numbers as one unit or is very familiar with them.

e. g. 'Beverley Hills 90210' (oh)

Because it has just one syllable as compared to the two-syllable 'zero', 'oh' is often used in the interest of speed and euphony. However, 'zero' is often clearer, especially on the phone, and leads to fewer misunderstandings.

2. Age and Counting Items

When referring to a person's age or when counting items, one will neither say 'oh' nor 'zero'.

e. g. "My great-grandmother is one hundred and two."

"There are a hundred and five lines in this text."

however

"Please refer to page one oh five." (p. 105)

Here the speaker is not counting, but simply referring to a single item.

b) The word **nought** is sometimes used before decimals 0.004.

This occurs only in British English. An American would not understand "nought".

c) We say **nil/nothing** in sports

five nil (UK)

five nothing (US)

In tennis, we say **love**, however, for zero. ("Love" supposedly originates from the French l'oeuf.)

Nil can also refer to zero, as in

"The chances of a first-year student getting into this class are nil."

NB (Nota Bene)

Oh and zero are usually interchangeable.

Sometimes expressed as nought in the UK before the decimal point.

German null translates as zero and not null, which is restricted to certain expressions like null and void (nichtig) or the null set (Nullmenge).

1.5.3 The Decimal Point

To separate thousands, millions and so on, we use a **comma** (,).

25,000 = twenty-five thousand

but a dot (.) is used and not a comma (,) for decimals in English.

25.000 = twenty-five point oh oh oh

0.005 = zero point oh oh five

1.5.4 Plural Forms and False Friends

a) plural forms, **no /s/**

There is no /s/ on million, billion, etc.

e. g. 82,000,000 = eighty-two million

exception → for large, non-specific, uncertain quantities, e. g. many, countless, etc. in conjunction with of

e. g. "Countless millions of euros are wasted in Brussels yearly."

"The star is thousands of light years away."

"Many hundreds of thousands of people died in the tsunami in Asia."

NB

This does not apply to a few and several. Here there is no /s/.

e. g. "The US budget amounts to several trillion dollars."

b) False Friends (billion/Billion)

How would you translate ten billion into German?

How would you translate zehn Billionen into English?

1.5.5 Dates

a) In modern English, it is customary to express dates differently than in German.
e.g. 1985 (year) nineteen eighty-five
and not nineteen hundred and eighty-five (antiquated)

(This is likely different from what you learned in school!)

b) With the digit 0 in the year, we say "oh".
1402 (year) fourteen oh two

c) The new millennium

Will we continue to use "two thousand and..." or revert to the original paradigm of "twenty..."?

1999 (year)	nineteen ninety-nine
2000 (year)	two thousand
2009 (year)	two thousand and nine
2020 (year)	two thousand and twenty
	OR twenty twenty

1.5.6 Fractions

Ordinal numbers (Ordinalzahlen) are usually used, with the exception of a half and a quarter. How would you read the following?

1/2

2/3

3/4

4/5

7/9 [seven-ninths]

When the denominator reaches a certain point, the fraction will be read differently.

1 = numerator

3 = denominator

e.g. $\frac{5}{212}$ five over two hundred and twelve

1.5.7 Percent

It can be written as one word **percent** or two **per cent**.

Make sure that the tonic accent falls on the second syllable, i. e. that the second syllable is stressed.

Otherwise, it sounds like person!

2 Wings and Airfoils

2.1 Geometry Definitions

2.1.1 Exercise A: Basic Vocabulary

A) *Match the letters in the following diagrams with the words provided.*

camber
leading edge
span
thickness
chord
trailing edge
tip
centreline
chord line
airfoil
aspect ratio
wing planform
mean camber line
dihedral angle
wing area

Top View (or A)

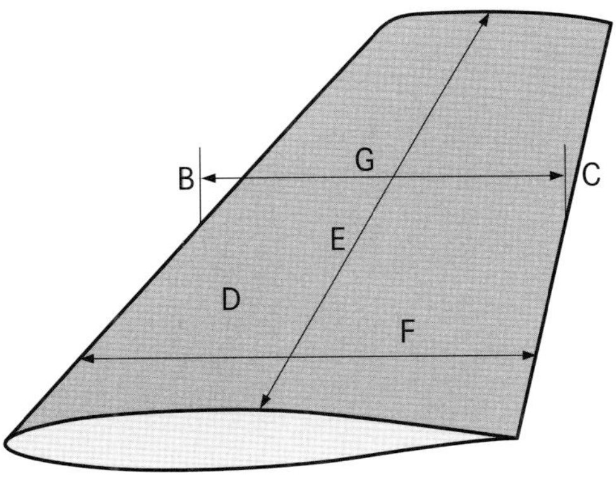

$$H = E^2 / D$$

Fig. 2.1 An Airfoil Seen from Above

Side View (or I)

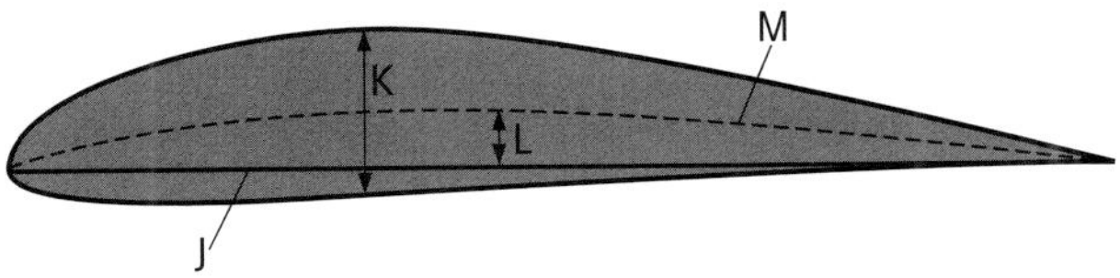

Fig. 2.2 Cross-sectional View of an Airfoil

Front View

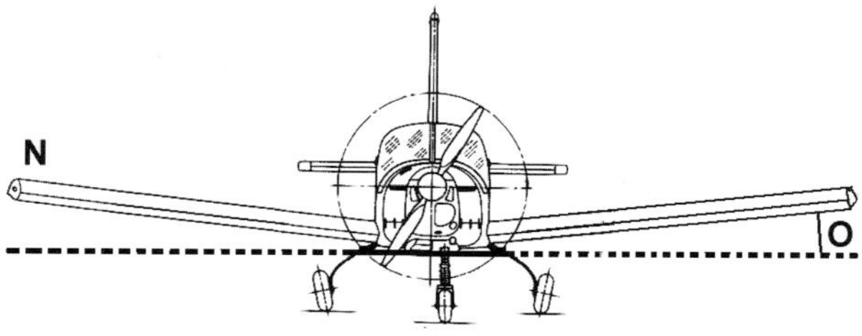

Fig. 2.3 Airfoils Seen from the Front of an Airplane

B) *Describe the significance of aspect ratio?*

Slower airplanes (.ex gliders : higher aspect ratio

C) What types of aircraft have a high aspect ratio? What types have a low aspect ratio?

2.1.2 Exercise B: Further Terminology

Label the diagram below with the following words:

D flap

B main airfoil

A slat

E slot

C vane

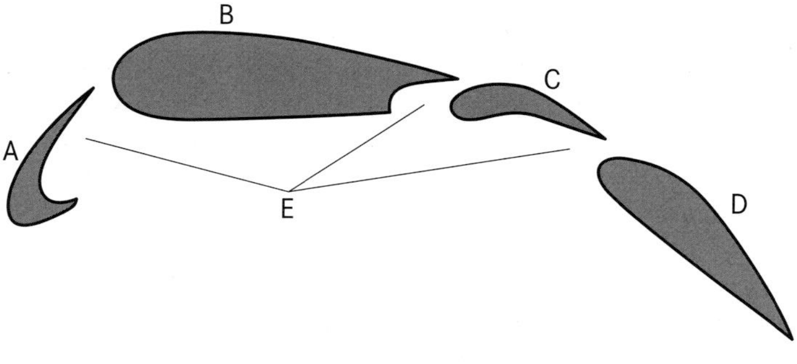

Fig. 2.4 Parts of an Airfoil

2.1.3 Exercise C: Swept Wings

Describe the aerodynamic advantages and disadvantages of a swept wing over a straight wing.

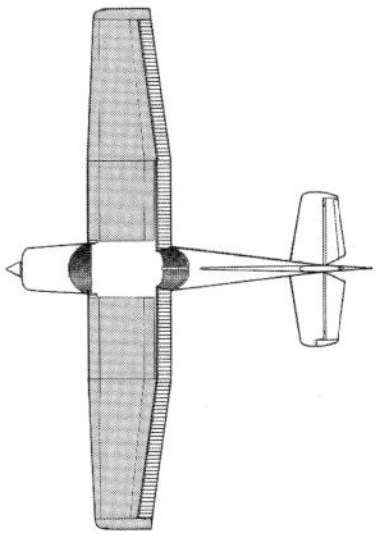

Fig. 2.5 Straight Wing

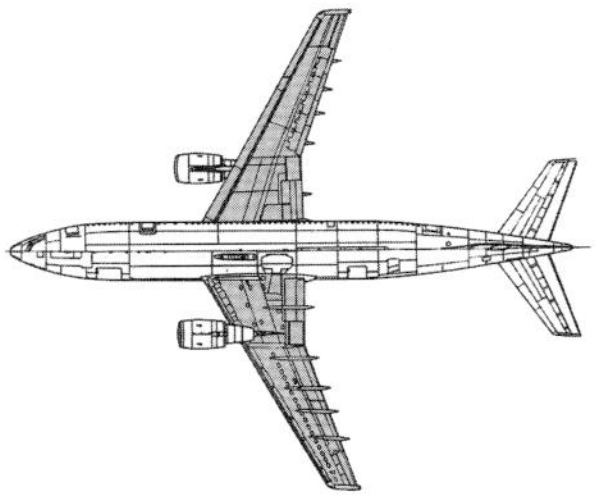

Fig. 2.6 Swept Wing

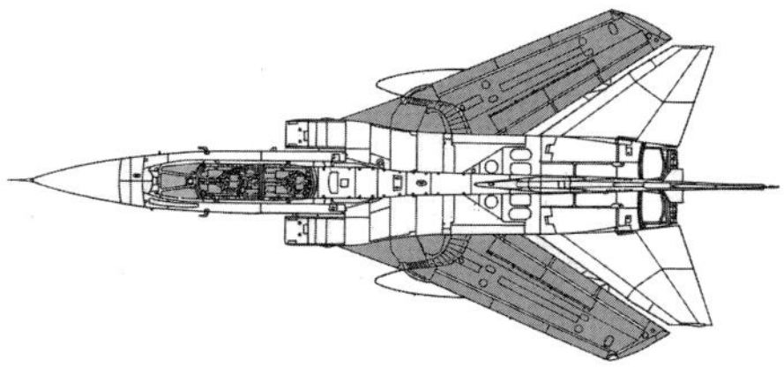

Fig. 2.7 Highly Swept Wing

2.2 The Secret Behind Mini-TEDs: Novel Aerodynamic Wing Flaps

1.
The fixed-wing geometry of a commercial aircraft only exhibits optimum aerodynamic properties within a relatively limited range of conditions, for example, when the aircraft is at its normal cruising altitude and speed. Outside this range, the aerodynamics will be less than optimal.

2.
If the wing geometry is designed for a broad performance spectrum, on the other hand, this detracts from the optimum performance in any specific area. For instance, high speeds require a thin, less curved wing, while short take-offs and landings require a thick wing with a highly cambered profile. The extendable/retractable slats and flaps on the leading and trailing edges of the wing, which have long been employed to provide sufficient lift at low speeds–for instance during take-off and landing–represent a sound compromise. However, aerodynamic research is tending towards the development of an adaptive wing that can constantly adjust its geometry to accommodate given flight conditions. Here too, however, engineers are forced to make compromises in terms of what is technically possible and economically reasonable. For this reason, existing flaps and spoilers are given expanded functionality while new, auxiliary systems with the simplest possible design are

integrated into the aircraft's overall system. The aerodynamics of the wing's trailing edge is a particularly important factor in the development of adaptive wing elements. Surprisingly, it was from Formula 1 cars that the decisive impetus came.

3.
In 1971, American Formula 1 driver and designer Dan Gurney discovered that a narrow strip of material bolted to the rear edge of a racing car's rear wing at a 90-degree angle to the airflow provided additional downforce, which translated into improved grip and higher speeds in bends. Former aircraft manufacturer McDonnell Douglas provided a wind tunnel for further investigation of the phenomenon, with the result that the Gurney Flap was used not only on racing cars but also on the DC-10 airliner and its successor model, the MD-11. 'Gurneys' were even installed on the tail stabilisers of certain helicopters. In these cases, though, the aim was generally to increase lift rather than downforce, and the strips were accordingly attached to the underside of the wing's trailing edge.

4.
Ten years ago, the 'Highlift concept' project funded by the German Federal Ministry of Economics developed an adjustable version of the basic, right-angled Gurney flap, thus producing the multifunction trailing edge devices known as 'mini-TEDs'. Within the scope of another multidisciplinary technology programme, Airbus worked with the EADS Corporate Research Centre to construct mini-TEDs and subject them to structural analysis on an early test-rig. These mini-TEDs are 9- to 15-centimetre-wide carbon fibre composite flaps that are integrated into the underside of the landing flap.

5.
In the trailing-edge airflow behind the extended mini-TEDs, two counter-rotating vortices form a slipstream that redirects the airflow downwards and produces greater lift. Far from disrupting the airflow over the wing surface, this effect actually has a beneficial influence. The landing flaps are able to generate even greater lift, because the mini-TEDs increase the critical angle of attack at which a stall would occur. The resulting drag remains within acceptable limits, and in some situations–such as landing approach–is even desirable. The increased lift generated by the mini-TEDs enables the aircraft to leave the ground sooner during take-off and use a steeper angle of approach when landing. This reduces the area overflown at low altitude, in turn narrowing the aircraft's noise footprint (defined for a noise level of 80 decibels, for example, which corresponds to urban traffic noise).

6.
The mini-TEDs can also have a positive impact when cruising at higher speeds, because the strength requirements for the wing–and therefore its weight–are determined in part by the stress due to turbulence. The mini-TEDs, which operate automatically and can deflect downwards rapidly, are linked via the flight control system to a turbulence sensor and can modify the lift distribution across the entire wingspan in a way that alleviates the load caused by gusts. If this mini-TED effect is taken into consideration at the airfoil design stage, the wing can be made lighter. A second economic plus is that at high airspeeds just under Mach 1, localized supersonic areas that form on the upper surface of the wing are concluded by a shock wave. This produces aerodynamic drag, which can be controlled and reduced by selectively redistributing lift through the deployment of mini-TEDs. A lighter aircraft with less drag consumes less fuel, reducing operating costs as well as pollutant emissions.

(adapted from Planet AeroSpace 3 / 2006)

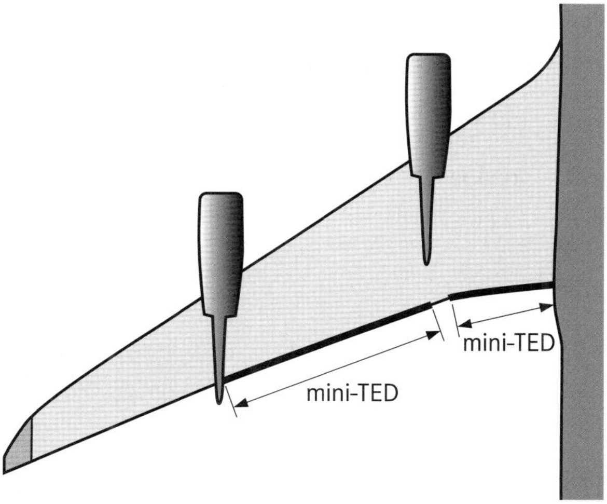

Fig. 2.8 Mini-TEDs Mounted on the Trailing Edge of an Airfoil

2.2.1 Exercise A: Comprehension Questions

After reading the text above on mini-TEDs, answer the following questions using full sentences and your own words as much as possible.

1) What are mini-TEDs?

Siehe 4. letzter Satz

- movable flaps
- carbon fibre
- multifunctional
- 9-15 cm
- underside of the landing flap

2) What benefits do they bring to commercial aircraft design?

- steeper angle of approach → smaller noise footprint
- lighter wing → less fuel → costs reduced
  → less emissions
- airflow is redirected downwards → greater lift

3) Explain how they work.

Behind the movable mini-TEDs two counter-rotating vortices are created. These form a slipstream due to which the airflow is redirected downwards and creates greater lift.

- react swiftly
- operated automaticly & rapidly

4) The text mentions the term 'broad performance spectrum' (paragraph 2). What is meant by this? Give examples from the text that would fit into this spectrum.

take-off, landing, ...

2.2.2 Exercise B: Technical Vocabulary

What is meant by the following technical terms in the text? The number of the paragraph where the word or expression is found in the text is provided in brackets.

Technical Term	Explanation / Definition
highly cambered profile (paragraph 2)	highly curved profile
retractable slats and flaps (2)	retract = to pull in (einfahrbar)
spoilers (2)	fixed flaps
test-rig (4) ↑ = Plattform	Versuchsstand = a platform where sth. is tested
slipstream (5) = Windschatten (nicht hier), Luftschraubenstrahl (hier)	hier: = a vortex of air after the airfoil
stall (5)	
noise footprint (5)	
gusts (6)	Böen = sudden and strong airstream
pollutant emission (6)	

2.2.3 Exercise C: Academic Vocabulary

Explain what the following terms mean in simpler English. The number of the paragraph where the word or expression is found in the text is provided in brackets.

Academic Vocabulary	Plain English
exhibits optimum aerodynamic properties (1)	shows
this **detracts** from the optimum performance (2)	takes away
which have long been **employed** (2)	used
a **sound** compromise (2)	good / reasonable
to **accommodate** given flight conditions (2)	adjust to / fit to
auxiliary systems (2)	(helping) / support
decisive impetus (2) ↳ related to "decision"	important / critical
its **successor** model (3)	next (sucessor <=> predecessor)
within the **scope** of another multidisciplinary technology programme (4)	area
to construct mini-TEDs and **subject them to** structural analysis (4) aussetzen/unterziehen	put them through
in a way that **alleviates** the load (6)	decrease / lighten
the **deployment** of mini-TEDs (6)	use

Fig. 2.9 Blended Wing Body Aircraft – A potential new aircraft design

2.2.4 Exercise D: Group Work

In groups of three or four, brainstorm other new design features for or adaptations to existing aircraft designs that would lead to lower pollutant emissions and/or noise reduction.

Rate the features you discuss in terms of efficiency and viability on a scale of 1 to 5 (with one being the least efficient and viable). What savings are involved and how likely are these features to be developed.

Be prepared to present the results of your discussion to the class.

Feature	Efficiency	Viability

2.3 Reading Mathematical Equations Aloud

This exercise can be done individually, in pairs or in groups.

2.3.1 Basic Arithmetic Operations

= ≙ equals / is equal to

How would you read the following equations and values?

$x + y = z$ x plus y equals z

$x - y = z$ x minus y equals z (subtract / take away /)

$a \times b = c$ a times / multiplied by b equals c

$a / b = c$ a divided by (over) b equals c

Name the four arithmetic operations above.

addition, subtraction, multiplication, division

2.3.2 Superscripts and Subscripts

Exponents are a type of superscript; that is they are written slightly above the preceding letter or number.

How would you read the following values?

$a^2 + b^2 = c^2$ a squared plus b squared equals c squared

a^3 a cubed

a^{-n} a to the power of minus n

a^{n-2} ....

Nota Bene (NB): In order to differentiate between a^{n-2} and a^n-2, a pause is used between the exponent "*n*" and "minus two" when reading the second expression. We do not normally insert brackets "( )" in English when they are not already written there.

Subscripts, which are written below the preceding letter or number, are usually read as "sub", the short form of subscript.

How would you read the following values?

$nx_1 - mx_2$ n times x sub one minus m times x sub two

Subscripts are also found in functions

$f_2 f_3 = f_2 [f_3 (x)]$ *fsubtwo times fsubthree equals fsubtwo of fsubthree of x*

2.3.3 Roots

How would you read the following values?

$\sqrt[2]{a} + \sqrt[3]{b} = c$ the square root of a plus the cube root of b equals c

$\sqrt[4]{a}$ the fourth root of a

$\sqrt[n]{\ }$ the nth root of

2.3.4 Brackets

In English we can translate *Klammern* "( )"with either parentheses, brackets or even round brackets. Generally, however, the expression brackets is used most predominantly in the English-speaking world. Other types of brackets include: square brackets "[ ]" and braces "{}", although the latter is usually used to denote a set and is therefore not read aloud.

How would you read the following values?

$(x + y)^2 (a - c)^3 = z$ x plus y in brackets squared times a minus c in brackets cubed equals z

$b = [(a + c) - (d + e)]$ ....

$\{2, 4, 6, 8\}$ the set two four six eight

2.4 Exercises

2.4.1 Further Mathematical Symbols

Fill in the German translation of each of the mathematical symbols below.

Symbol	Reading	German Translation
±	plus minus (plus or minus)	plus minus
>	greater than	größer als
<	less than	kleiner als
≥	greater than or equal to	größer gleich
≤	less than or equal to	kleiner gleich
≈	approximately equal to	ungefähr
≠	is not equal to	ungleich
→	approaches	gegen
∴	therefore	daraus folgt (=>)
∞	infinity	unendlich

2.4.2 Further Practice

Write out the following equations, including all numbers, in words.

1) $x^{-p} = 1/x^p$

2) $x = \dfrac{n x_1 + m x_2}{m + n}$

3) $y \neq 10{,}001$

4) $b^2 \geq a^2 (1 - e^2)$

5) $\sqrt[3]{x} + \sqrt[n]{6} = y$

3 Aircraft Motion

There are three main movements of an airplane, each acting along one of the three-dimensional axes: pitch, roll and yaw. Each movement requires the use of certain parts of an aircraft.

3.1 Pitch Motion

Read the text below and then label the figure with the pitch axis, pitch motion, centre of gravity and the parts of the airplane involved in the pitch motion.

Fig. 3.1 Parts of an Airplane Used for Pitch Motion

1.
In flight, any aircraft will rotate about its centre of gravity, a point which is the average location of the mass of the aircraft. We can define a three-dimensional coordinate system through the centre of gravity with each axis of this coordinate system perpendicular to the other two axes. We can then define the orientation of the aircraft by the amount of rotation of the parts of the aircraft along these principal axes. The **pitch axis** is perpendicular to the aircraft centreline and lies in the plane of the wings. A **pitch motion** is an up or down movement of the nose of an aircraft.

2.
The pitching motion is caused by the deflection of the elevator of an aircraft. The elevator is a hinged section at the rear of the horizontal stabilizer. There is usually an elevator on each side of the vertical stabilizer. The elevators work in pairs; when the right elevator goes up, the left elevator also goes up.

3.
Changing the angle of deflection at the rear of an airfoil changes the amount of lift generated by the foil. With greater downward deflection, lift increases in the upward direction. With greater upward deflection, lift increases in the downward direction. The lift generated by the elevator acts through the centre of pressure of the elevator and horizontal stabilizer and is located at some distance from the centre of gravity of the aircraft. The change in lift created by deflecting the elevator generates a torque about the centre of gravity which causes the airplane to rotate. The pilot can use this ability to make the airplane loop. Or, since many aircraft loop naturally, the deflection can be used to trim or balance the aircraft, thus preventing a loop.

4.
On many aircraft the horizontal stabilizer and elevator create a symmetric airfoil. This produces no lift when the elevator is aligned with the stabilizer and allows the combination to produce either positive or negative lift, depending on the deflection of the elevator. On many fighter planes, in order to meet their high maneuvering requirements, the stabilizer and elevator are combined into one large moving surface called a stabilator. The change in force is created by changing the inclination of the entire surface, not by changing its effective shape.

(courtesy of NASA Glenn Research Center)

3.2 Roll Motion

Read the text below and then label the figure with the roll axis, roll motion, centre of gravity and the parts of the airplane involved in roll motion.

Fig. 3.2 Parts of an Airplane Used for Roll Motion

1.
In flight, any aircraft will rotate about its centre of gravity, a point which is the average location of the mass of the aircraft. We can define a three-dimensional coordinate system through the centre of gravity with each axis of this coordinate system perpendicular to the

other two axes. We can then define the orientation of the aircraft by the amount of rotation of the parts of the aircraft along these principal axes. The **roll axis** lies along the aircraft centreline. A **roll motion** is an up and down movement of the wings of an aircraft.

2.
The rolling motion is caused by the deflection of the ailerons of this aircraft. An aileron is a hinged section at the rear of each wing. The ailerons work in opposition; when the right aileron goes up, the left aileron goes down.

3.
Changing the angle of deflection at the rear of an airfoil will change the amount of lift generated by the foil. With greater downward deflection, the lift will increase in the upward direction; with greater upward deflection, the lift will decrease in the upward direction. Since the ailerons work in pairs, the lift on one wing increases as the lift on the opposite wing decreases. Because the forces are not equal, there is a net twist, or torque about the centre of gravity and the aircraft rotates about the roll axis. The pilot can use this ability to bank the aircraft which causes the airplane to turn.

4.
On this page we have mentioned how an aircraft roll is induced by the movement of the ailerons, but there are other ways to produce a rolling motion on an aircraft. The Wright brothers used a method called wing warping. Their wings were wired together in such a way that the outer panels of each wing could be twisted relative to the inner panel. The twisting changed the local angle of attack of sections of the wing which changed the lift being generated by that section. Unequal forces on the wings caused the aircraft to roll. Many modern airliners use a spoiler to roll the aircraft. A spoiler is a plate that is raised between the leading and trailing edges of the wing. The spoiler effectively changes the shape of the airfoil, disrupts the flow over the wing, and causes a section of the wing to decrease its lift. This produces an unbalanced force with the other wing, which causes the roll. Airliners use spoilers because spoilers can react more quickly than ailerons and require less force to activate, but they always decrease the total amount of lift for the aircraft. You can tell whether an airliner is using spoilers or ailerons by noticing where the moving part is located. At the trailing edge, it's an aileron; between the leading and trailing edges, it's a spoiler.

(courtesy of NASA Glenn Research Center)

3.3 Yaw Motion

Read the text below and then label the figure with the yaw axis, yaw motion, centre of gravity and the parts of the airplane involved in yaw motion.

1.
In flight, any aircraft will rotate about its centre of gravity, a point which is the average location of the mass of the aircraft. We can define a three-dimensional coordinate system through the centre of gravity with each axis of this coordinate system perpendicular to the other two axes. We can then define the orientation of the aircraft by the amount of rotation of the parts of the aircraft along these principal axes. The **yaw axis** is perpendicular to the wings and lies in the plane of the aircraft centreline. A **yaw motion** is a side-to-side movement of the nose of an aircraft.

Fig. 3.3 Parts of an Airplane Used for Yaw Motion

2.
The yawing motion is caused by the deflection of the rudder of this aircraft. The rudder is a hinged section at the rear of the vertical stabilizer.

3.
Changing the angle of deflection at the rear of an airfoil changes the amount of lift generated by the foil. For the vertical stabilizer and rudder, the orientation of the airfoil causes a side force to be generated. With greater deflection of the rudder to the left, the side force increases to the right. With greater deflection to the right, the side force increases to the left. The lift generated by the rudder acts through the centre of pressure of the rudder and vertical stabilizer and is located at some distance from the centre of gravity of the aircraft. The change in side force created by deflecting the rudder generates a torque about the centre of gravity, which causes the airplane to rotate. The pilot uses this ability to keep the nose of the aircraft pointed in the direction of travel.

4.
On all aircraft, the vertical stabilizer and rudder create a symmetric airfoil. This produces no side force when the rudder is aligned with the stabilizer and allows the combination to produce either positive or negative side force, depending on the deflection of the rudder. Some fighter planes have two vertical stabilizers and rudders because of the need to control the plane with multiple, very powerful engines

(courtesy of NASA Glenn Research Center)

3.4 Trim Systems

Although an airplane can be operated throughout a wide range of attitudes, airspeeds, and power settings, it can only be designed to fly hands off within a very limited combination of these variables. Therefore, trim systems are used to relieve the pilot of the need to maintain constant pressure on the flight controls. Trim systems usually consist of cockpit controls and small hinged devices attached to the trailing edge of one or more of the primary flight control surfaces. They are designed to help minimize a pilot's workload by aerodynamically assisting movement and position of the flight control surface to which they are attached. Common types of trim systems include trim tabs, balance tabs, antiservo tabs, ground adjustable tabs, and an adjustable stabilizer.

Trim Tabs

The most common installation on small airplanes is a single trim tab attached to the trailing edge of the elevator. Most trim tabs are manually operated by a small, vertically mounted control wheel. However, a trim crank may be found in some airplanes. The cockpit control includes a tab position indicator. Placing the trim control in the full nose-down position moves the tab to its full up position. With the tab aimed up and into the airstream, the airflow over the horizontal tail surface tends to force the trailing edge of the elevator down. This causes the tail of the airplane to move up, and results in a nose-down pitch change (Fig. 3.4).

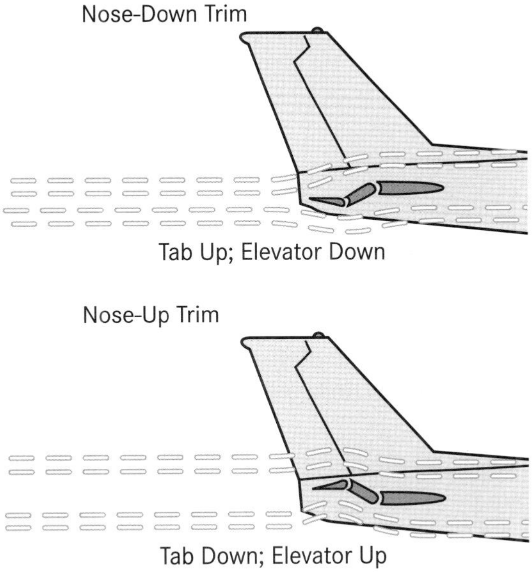

Fig. 3.4 The movement of the elevator is opposite to the direction of movement of the elevator trim tab

If you set the trim tab to the full nose-up position, the tab moves to its full-down position. In this case, the air flowing under the horizontal tail surface hits the tab and tends to force the trailing edge of the elevator up, reducing the elevator's angle of attack. This causes a tail-down movement of the airplane and a nose-up pitch change.

In spite of the opposite direction movement of the trim tab and the elevator, controlling trim is natural to a pilot. If you have to exert constant back pressure on the control column, the need for nose-up trim is indicated. The normal trim procedure is to continue trimming until the airplane is balanced and the nose-heavy condition is no longer apparent. Pilots normally establish the desired power, pitch attitude, and configuration first, and then trim the airplane to relieve control pressures that may exist for that flight condition. Any time power, pitch attitude, or configuration is changed, expect that retrimming will be necessary to relieve the control pressures for the new flight condition.

Balance Tabs

The control forces may be excessively high in some airplanes, and in order to decrease them, the manufacturer may use balance tabs. They look like trim tabs and are hinged in approximately the same places as trim tabs. The essential difference between the two is that the balancing tab is coupled to the control surface rod so that when the primary control surface is moved in any direction, the tab automatically moves in the opposite direction. In this manner, the airflow striking the tab counter-balances some of the air pressure against the primary control surface, and enables the pilot to more easily move and hold the control surface in position.

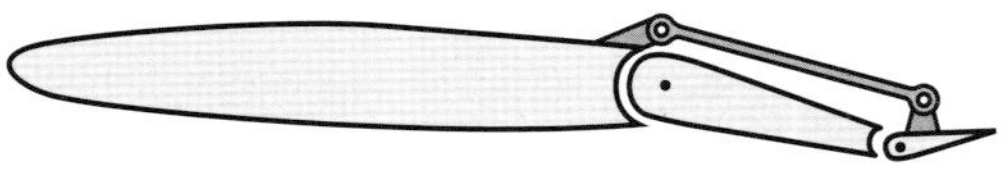

Fig. 3.5 Illustration of a Balance Tab

If the linkage between the tab and the fixed surface is adjustable from the cockpit, the tab acts as a combination trim and balance tab, which can be adjusted to any desired deflection. Any time the control surface is deflected, the tab moves in the opposite direction and eases the load on the pilot.

Antiservo Tabs

In addition to decreasing the sensitivity of the stabilator, an antiservo tab also functions as a trim device to relieve control pressure and maintain the stabilator in the desired position. The fixed end of the linkage is on the opposite side of the surface from the horn on the tab, and when the trailing edge of the stabilator moves up, the linkage forces the trailing edge of the tab up. When the stabilator moves down, the tab also moves down. This is different than trim tabs on elevators, which move in opposition to the control surface (Fig. 3.6).

This tab works in the same manner as the balance tab except that, instead of moving in the opposite direction, it moves in the same direction as the trailing edge of the stabilator. For example, when the trailing edge of the stabilator moves up, the linkage forces the trailing edge of the tab up. When the stabilator moves down, the tab also moves down.

Ground Adjustable Tabs

Many small airplanes have a non-movable metal trim tab on the rudder. This tab is bent in one direction or the other while on the ground to apply a trim force to the rudder. The correct displacement is determined by a trial-and-error process. Usually, small adjustments are necessary until you are satisfied that the airplane is no longer skidding left or right during normal cruising flight (Fig. 3.7).

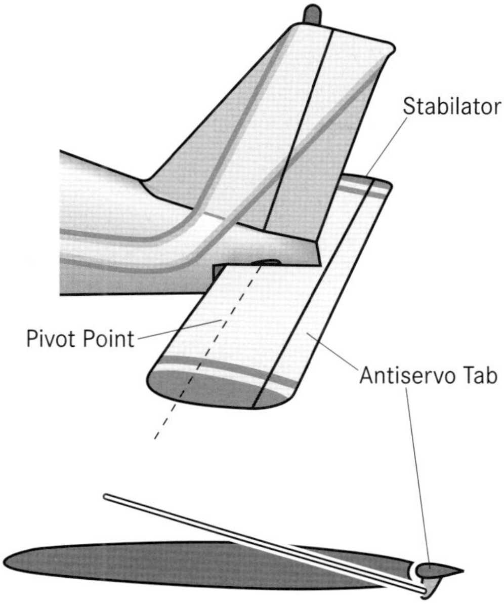

Fig. 3.6 An antiservo tab attempts to streamline the control surface and is used to make the stabilator less sensitive by opposing the force exerted by the pilot

Fig. 3.7 Ground Adjustable Tab on a Cessna 172

Adjustable Stabilizer

Rather than using a movable tab on the trailing edge of the elevator, some airplanes have an adjustable stabilizer. With this arrangement, linkages pivot the horizontal stabilizer about its rear spar. This is accomplished by use of a jackscrew mounted on the leading edge of the stabilator (Fig. 3.8).

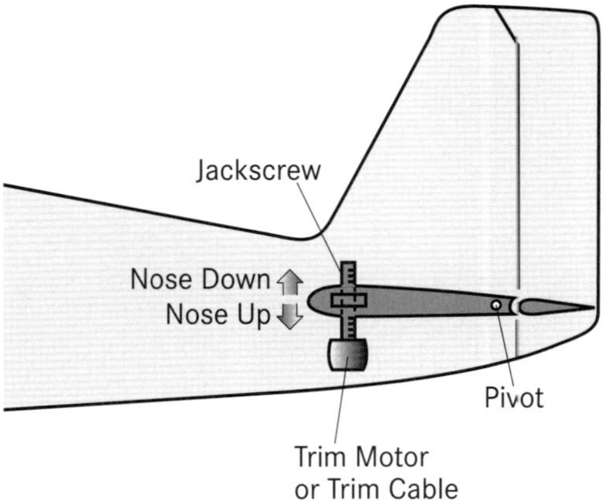

Fig. 3.8 Some airplanes, including most jet transports, use an adjustable stabilizer to provide the required pitch trim forces

On small airplanes, the jackscrew is cable-operated with a trim wheel or crank, and on larger airplanes, it is motor driven. The trimming effect and cockpit indications for an adjustable stabilizer are similar to those of a trim tab.

(adapted from Pilot's Handbook of Aeronautical Knowledge, FAA; 4-8, 4-9)

3.4.1 Comprehension Questions

1) *Provide a definition of trimming?*

....

....

....

2) *Why are trim systems useful?*

....

....

....

3) *Explain Figure 3.4 in your own words.*

....

....

....

4) *What is the purpose of an antiservo tab and ground adjustable tabs?*

3.4.2 Vocabulary

Provide the German equivalent of the words below taken from the text.

English Term	German Equivalent
crank (excerpt on trim tabs) crank shaft	Kurbel Kurbelwelle
hinged (excerpt on balance tabs) a hinge	gelenkig Gelenk / Scharnier
rod (balance tabs)	Stange
skidding (ground adjustable tabs)	rutschen / schleudern
pivot (adjustable atabilizer) a pivot	drehen Drehpunkt / Drehachse
rear spar (adjustable stabilizer)	Hinterholm
jackscrew (adjustable stabilizer) a screw to jack sth. up = to lift sth.	Schraube

3.5 Mathematical Symbols and Formulae

a) Match the symbols and formulae with the expressions provided below. In some cases you need to assign more than one expression to each symbol or expression.

b) Once you have finished assigning the expressions, write out the complete reading of each symbol or formula.

1)	$\lvert x \rvert$	2)	$n\,!$
3)	$\lim_{x \to \infty} f(x)$	4)	$\int_a^b f(x)\,dx$
5)	$\sum_{d=1}^{n} B_d$	6)	$\emptyset$
7)	$a \propto b$	8)	d/dx
9)	$3x / x = 3$	10)	$A \cap B$

1) the absolute value of x
2) n factorial OR factorial n
3) the limit of f of x as x approaches infinity
4) the definite integral of f of x from a to b
5) the sum of as d runs from one to n of Bsub d
6) the null set
7) a is proportional to b
8) the first derivative with respect to x
9) the x is cancelled
10) the intersection of A and B

null set	absolute value	intersection
cancel	derivative	proportional to
with respect to	the sum of	approaches
factorial	definite integral	limit of

4 Propulsion Systems

There are four main types of propulsion systems used in planes and spacecraft and they include: propeller propulsion, gas turbine propulsion, ramjet propulsion and rocket propulsion.

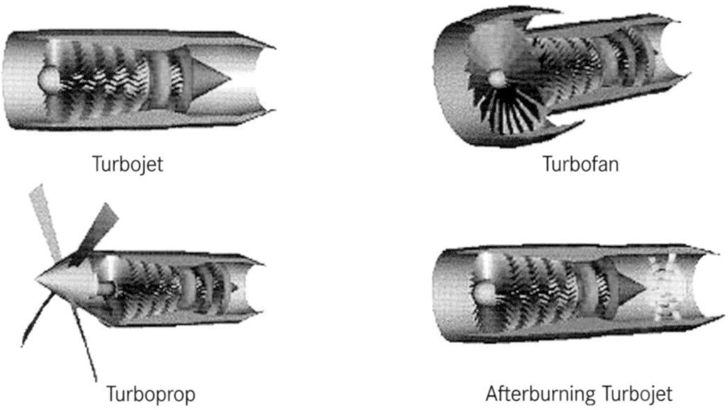

Fig. 4.1 Various Types of Propulsion Systems

Read about these four propulsion systems and then answer the questions after each text.

4.1 Propeller Propulsion

Propeller-Produced Thrust

For the forty years following the first flight of the Wright brothers, airplanes used internal combustion engines to turn propellers to generate thrust. Today, most general aviation or private airplanes are still powered by propellers and internal combustion engines, much like an automobile engine. The engine takes air from the surroundings, mixes it with fuel, burns the fuel to release the energy in the fuel, and uses the heated gas exhaust to move a piston which is attached to a crankshaft. In an automobile, the shaft is used to turn the wheels of the car. In an airplane, the shaft is connected to a propeller.

Fig. 4.2 Wright Flyer

Fig. 4.3 Do 228

Propellers as Airfoils

The details of propeller propulsion are very complex, but we can learn some of the fundamentals by using a simple momentum theory. The details are complex because the propeller acts like a rotating wing creating a lift force by moving through the air. For a propeller-powered aircraft, the gas that is accelerated, or the working fluid, is the surrounding air that passes through the propeller. The air that is used for combustion in the engine provides very little thrust. Propellers can have several blades, which are usually long and thin. A cut through the blade perpendicular to the long dimension will give an airfoil shape. Because the blades rotate, the tips move faster than the hub. So, to make the propeller efficient, the blades are usually twisted from hub to tip. The angle of attack of the airfoils at the tips is lower than at the hub.

Fig. 4.4 Four-bladed Propeller

Fig. 4.5 Eight-bladed Propeller

The Engines that Drive Propellers

The engine used in the P-51 was an internal combustion engine. After World War II, as jet engines gained popularity, aerodynamicists used jet engines to turn the propellers on some aircraft. This propulsion system is called a turboprop. The Airbus A400M transport plane is a turboprop aircraft. Its main thrust comes from the propellers, but the propellers are turned by turbine engines. The human-powered aircraft of the mid-80s were also propeller-powered, but the "engine" was provided by a human using a bicycle gearing device. Currently NASA is flying a solar-powered, electric engine aircraft that also uses propellers. Propeller-powered aircraft are very efficient for low-speed flight. But as the speed of the aircraft increases, regions of supersonic flow, with associated performance losses due to shock waves, occur on the propeller. Propellers are not used on high-speed aircraft.

(courtesy of Glenn Research Center, www.grc.nasa.gov)

Fig. 4.6 P-51 Mustang

Fig. 4.7 Airbus A400M

1) *Translate the following words taken from the text into German.*

- internal combustion engine Verbrennungsmotor
- piston Kolben
- crankshaft Kurbelwelle

2) *What is the working fluid in a propeller-powered aircraft?*

The working fluid is the surrounding air that passes through the propeller.

3) *Why are propeller blades usually twisted from the hub to the tips?*

(*) the blades are twisted in order to make the propeller efficient.

(*) The tips move faster than the hub. Therefore

4) *What is a turboprop?*

A turboprop is a propulsion system whose main thrust comes from the propellars which are turned by turbine engines.

4.2 Gas Turbine Propulsion

During World War II, a new type of airplane engine was developed independently in Germany and in England. This engine was called a gas turbine engine. We sometimes call this engine a jet engine. Early gas turbine engines worked much like a rocket engine creating a hot exhaust gas which was passed through a nozzle to produce thrust. But unlike the rocket engine which must carry its oxygen for combustion, the turbine engine gets its oxygen from the surrounding air. A turbine engine does not work in outer space because there is no surrounding air. For a gas turbine engine, the accelerated gas, or working fluid, is the jet exhaust. Most of the mass of the jet exhaust comes from the surrounding atmosphere. Most modern, high-speed passenger and military aircraft are powered by gas turbine engines.

Turbine engines come in a wide variety of shapes and sizes because of the many different aircraft missions. All gas turbine engines have some parts in common, however. Each aircraft has a unique mission and therefore a unique propulsion requirement. A Boeing 777 airliner, for example, has to carry large loads of passengers or cargo for a long distance at high speed. It spends most of its life in high-speed cruise. An F-22 Raptor fighter plane, on the other hand, has to be able to shoot down other aircraft in air-to-air combat. It spends most of its life in cruise, but needs high acceleration when in combat. An Airbus A400M cargo aircraft, like a Boeing 777, carries cargo a long distance, but it does not have the same high-speed requirements. A T-38 trainer is used to teach pilots how to fly jet aircraft and does not have the acceleration requirements of an F-22 Raptor. A Boeing 777 is powered by two high-bypass turbofan engines, an F-22 Raptor by two afterburning low-bypass turbofans, an Airbus A400M by four turboprop engines and a T-38 by turbojet engines.

(courtesy of Glenn Research Center, www.grc.nasa.gov)

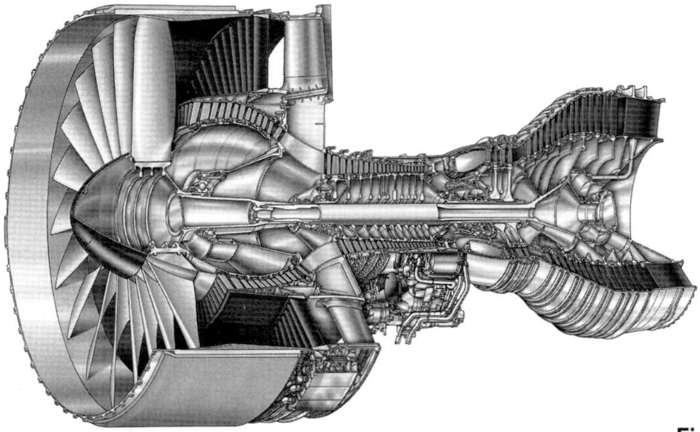

Fig. 4.8 Gas Turbine PW 4000

Comprehension and Vocabulary Questions

1) *How are jet engines similar to and different from rocket engines?*

....

....

....

....

2) *Translate the following words taken from the text in German:*

- nozzle
- exhaust gas
- high-speed cruise

3) *What is the difference between a Boeing 777 and an Airbus A400M?*

....

....

....

....

4.3 Ramjet Propulsion

In the early 1900s some of the original ideas concerning ramjet propulsion were first developed in Europe. In conventional engines, thrust is produced by passing the hot exhaust from the combustion of a fuel through a nozzle. The nozzle accelerates the flow, and the reaction to this acceleration produces thrust. To maintain the flow through the nozzle, the combustion must occur at a pressure that is higher than the pressure at the nozzle exit. In a ramjet, the high pressure is produced by "ramming" external air into the combustor using the forward speed of the vehicle. The external air that is brought into the propulsion system becomes the working fluid, much like a turbojet engine. In a turbojet engine, the high pressure in the combustor is generated by a piece of machinery called a compressor. But there are no compressors in a ramjet. Therefore, ramjets are lighter and simpler than a turbojet. Ramjets produce thrust only when the vehicle is already moving; ramjets cannot produce thrust when the engine is stationary or static. Since a ramjet cannot produce static thrust, some other propulsion system must be used to accelerate the vehicle to a speed where the ramjet begins to produce thrust. The higher the speed of the vehicle, the better a ramjet works until aerodynamic losses become a dominant factor.

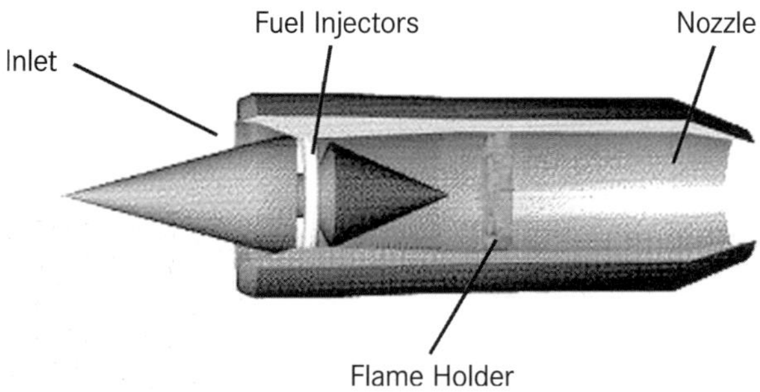

Fig. 4.9 A Ramjet Propulsion System

The combustion that produces thrust in the ramjet occurs at a subsonic speed in the combustor. For a vehicle travelling supersonically, the air entering the engine must be slowed to supersonic speeds by the aircraft inlet. Shock waves present in the inlet cause performance losses for the propulsion system. Above Mach 5, ramjet propulsion becomes very inefficient. The new supersonic combustion ramjet, or scramjet, solves this problem by performing the combustion supersonically in the burner.

Some applications include an X-15 rocket-powered airplane with a ramjet slung underneath the body and a missile with a ramjet. With both propulsion systems, a rocket is used to bring the ramjet up to speed before it produces thrust. Because the ramjet uses external air for combustion, it is a more efficient propulsion system for flight within the atmosphere than a rocket, which must carry all of its oxygen. Ramjets are ideally suited for very high-speed flight within the atmosphere.

(courtesy of Glenn Research Center, www.grc.nasa.gov)

Comprehension and Vocabulary Questions

1) *What are the differences between a ramjet and a turbojet?*

2) *Explain the origin of the term 'ramjet'?*

3) *What is the difference between 'subsonic', 'supersonic' and 'transsonic'? How would you translate these into German?*

4.4 Rocket Propulsion

Fig. 4.10 Ariane rocket engine

Fig. 4.11 Ariane 5

During and following World War II, there were a number of rocket-powered aircraft built to explore high-speed flight. The X-1A, used to break the sound barrier, and the X-15 were rocket-powered airplanes. In a rocket engine, fuel and a source of oxygen, called an oxidizer, are mixed and exploded in a combustion chamber. The combustion produces hot exhaust which is passed through a nozzle to accelerate the flow and produce thrust. For a rocket, the accelerated gas, or working fluid, is the hot exhaust produced during combustion. This is a different working fluid than you find in a turbine engine or a propeller powered aircraft. Turbine engines and propellers use air from the atmosphere as the working fluid, but rockets use the combustion exhaust gases. In outer space there is no atmosphere, so turbines and propellers cannot work there. This explains why a rocket works in space, whereas a turbine engine or propeller does not.

There are two main categories of rocket engines: liquid and solid rockets. In a liquid rocket, the propellants, the fuel and the oxidizer are stored separately as liquids and are pumped into the combustion chamber of the nozzle where burning occurs. In a solid rocket, the propellants are mixed together and packed into a solid cylinder. Under normal temperature conditions, the propellants do not burn, but they will burn when exposed to a source of heat provided by an igniter. Once the burning starts, it proceeds until all the propellant is exhausted. With a liquid rocket, you can stop the thrust by turning off the flow of propellants, but with a solid rocket, you have to destroy the casing to stop the engine. Liquid rockets tend to be heavier and more complex because of the pumps and storage tanks. The propellants are loaded into the rocket just before launch. A solid rocket is much easier to handle and can sit for years after firing.

The X-15 was powered by a liquid rocket engine and carried a single pilot to a height of more than 60 miles above the earth. The X-15 flew more than six times the speed of sound nearly 30 years ago. These altitude and speed records for a piloted aircraft are only exceeded today by the Space Shuttle.

(courtesy of Glenn Research Center, www.grc.nasa.gov)

Fig. 4.12 X-15

Comprehension and Vocabulary Questions

1) *How is the working fluid in a rocket engine different from that in a turbine engine?*

....

....

....

2) *Translate the following into German:*

- sound barrier
- exceeded (last sentence)
- propellant (2nd paragraph)

3) *Why is a liquid rocket more complex than a solid one?*

4.5 Paraphrasing

The following are some expressions dealing with paraphrasing (or summarizing in your own words).

Paraphrasing
The text is about ...
According to the text/author ...
Essentially, ...
Basically, the text is about ...
To put it another way ...
Putting it another way, ...
The basic idea of the text is ...
What the author is trying to say ...
In other words ...

Exercise: Paraphrasing

Find a partner and summarize in your own words one of the preceding texts dealing with propulsion systems (propeller, gas turbine, ramjet or rocket). Try not to look at your text while you are summarizing. Reverse the roles to ensure that both partners have had a turn at paraphrasing his or her own text.

4.6 Jet Engines

Working in groups of two or three, describe using full sentences how a jet engine works. Use the terms in the illustration of a jet turbine engine below in your description.

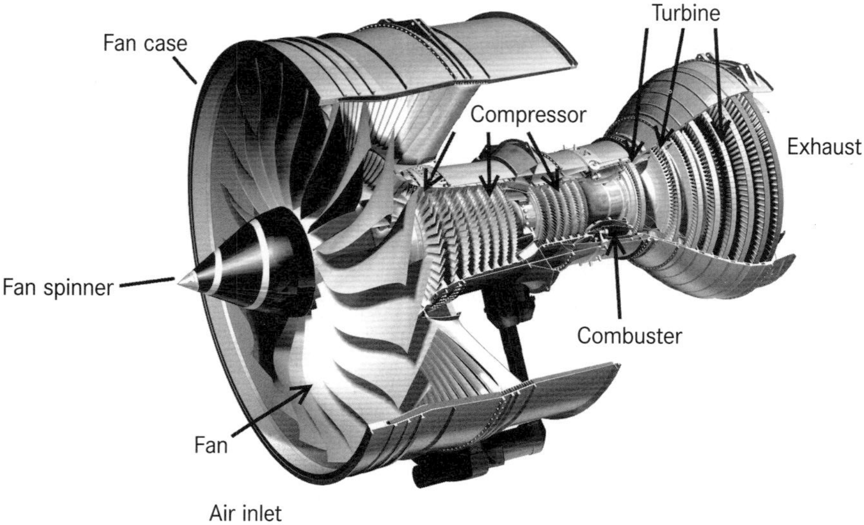

Fig. 4.13 Jet turbine engine

....

....

....

....

....

....

....

....

4.7 Shapes

As an aerospace engineer, you will be called upon from time to time to describe an instrument, device or object on the basis of its shape and appearance because either you or your colleague are unfamiliar with the object or you do not know its name.

Exercise: Naming Shapes

Label the shapes below and provide the adjective form for each one, e.g. triangle (noun), triangular (adjective).

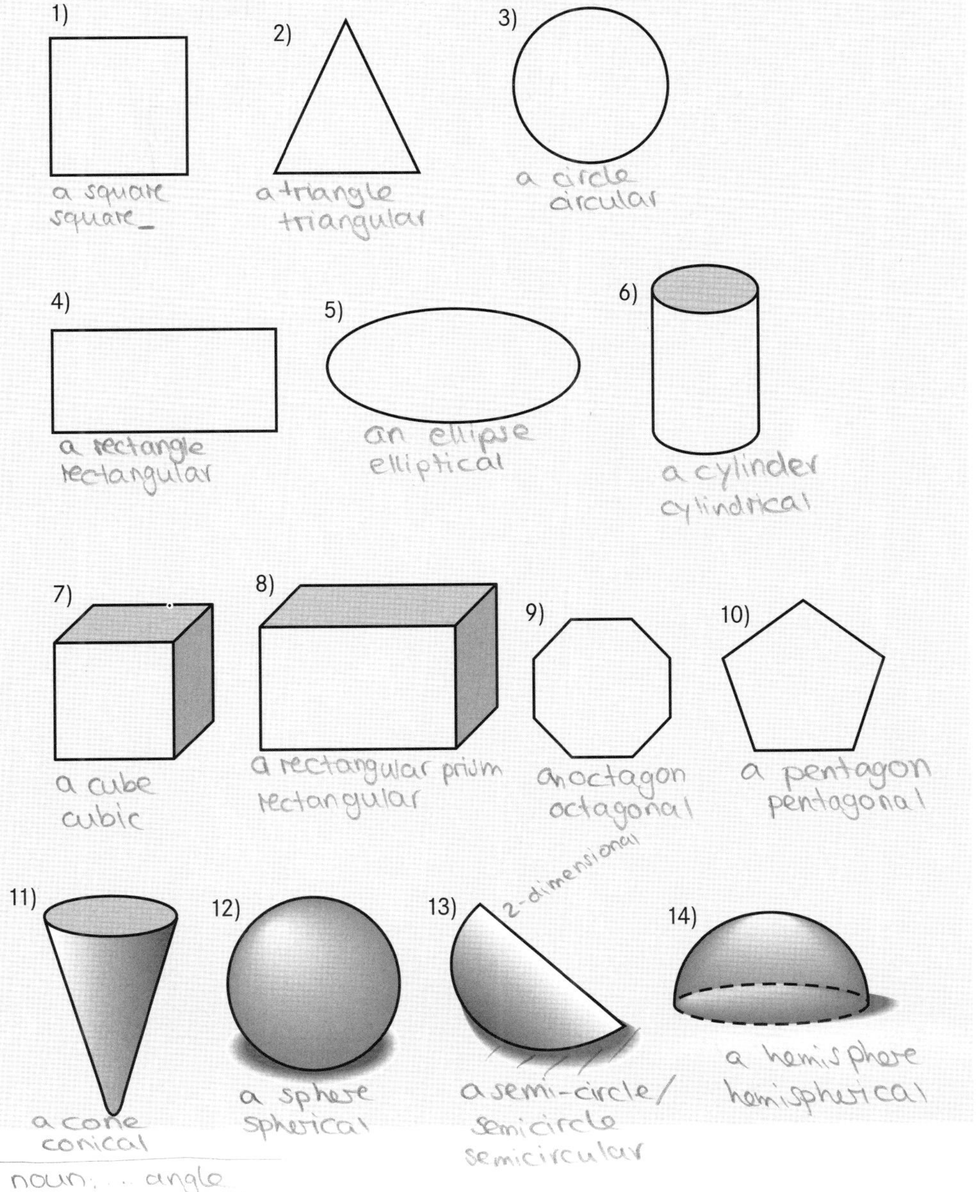

noun: angle
adjectiv: angular

4.8 Geometry

4.8.1 Exercise A

Match the following expressions with the pictures below.

acute angle	arc	area	height
bisector	chord	circumference	length
diameter	external angle	obtuse angle	width
parallel	perimeter	perpendicular to	breadth
radius	right angle		

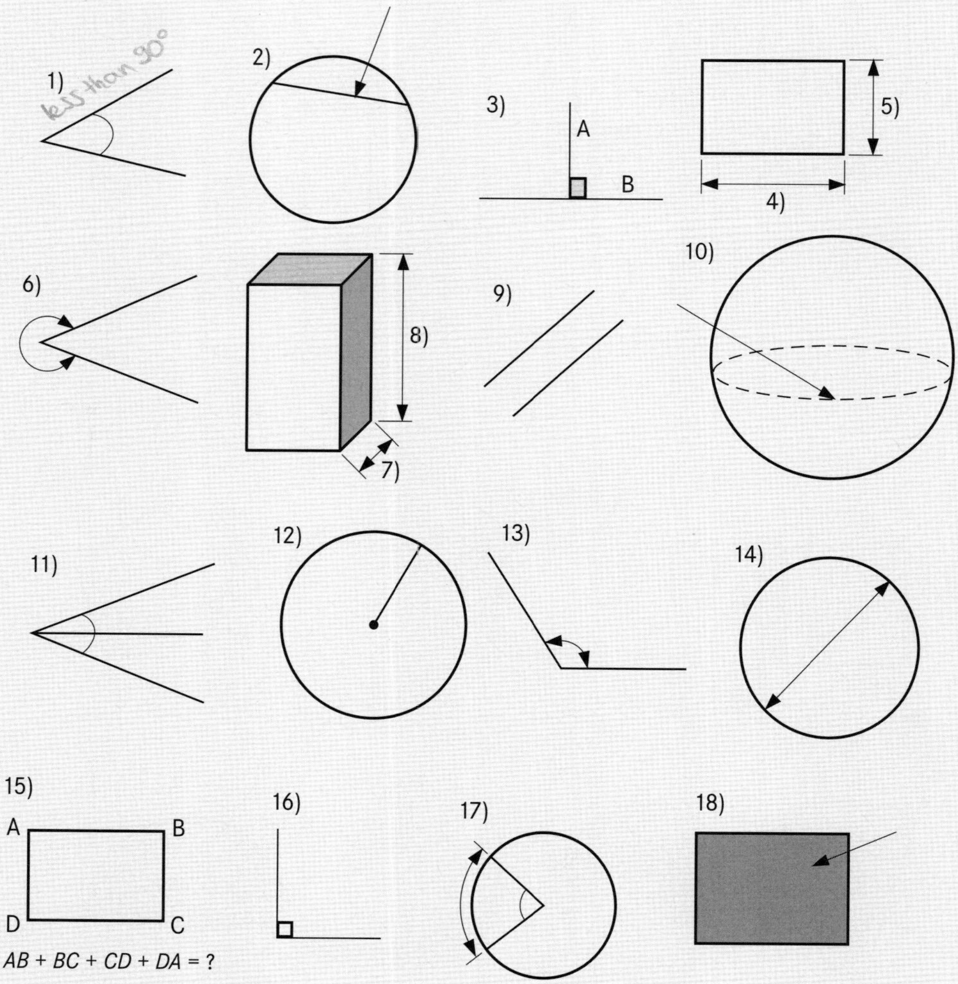

4.8.2 Exercise B

Describe the shape of a tool, gadget or machine to your partner and have your partner guess the object based on your description. You might want to make a few notes first.

....

....

....

....

....

....

....

....

5 Composite Materials

5.1 Vocabulary

Fill in the blanks with the words provided. Each word can be used only once and there are more words than blanks.

soft	reactive	inert	ductile
semi-conductive	malleable	thick	shiny
dull	tough	soluble	sratch-resistant
brittle	smooth	coarse	conductive
transparent	durable	translucent	pitted
matte	opaque	heat-resistant	corrosive
corrugated	corrosion-resistant	noble	composites

1. High carbon steel is less .... than mild carbon steel, which means it can't be shaped as easily.
2. Copper is often used in electric wiring because it is .... and ....
3. One of the reasons fuel cell vehicles have not made their way to mass production yet is because of safety issues. Hydrogen is a very .... gas.
4. Silicon and germanium are ....
5. A windowpane is a .... solid with a .... surface.
6. Cast iron is an example of a .... material; when it is subjected to stress it fractures.
7. Helium, neon, argon, krypton, xenon and radon are .... or .... gases.
8. Metals can be combined to form stronger, more heat and corrosion-resistant compounds called ....
9. The surface of copper becomes .... if not polished.

5.2 Composite Materials

5.2.1 Background Information

"Stopping the cracks", New Scientist 15 July 2006 (abridged and modified)

Composites are immensely popular engineering materials because they are up to 40 per cent lighter than aluminium yet 20 per cent stronger. However, great care must be taken when using them because composites fatigue and fail in a different way from metals.

Aviation composites comprise a polymer resin matrix that contains layers of strong reinforcing carbon or glass fibres. Layer after layer of these reinforced matrices are glued together and shaped in a curing oven to provide precisely the strength each component needs.

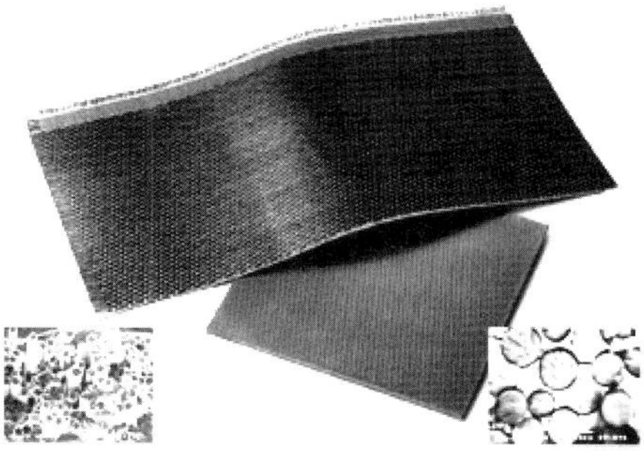

Fig. 5.1 Various Types of Composite Materials

With metals, although a fatigue crack can appear at a very low stress level, such cracks normally grow only slowly, allowing ample time to spot them before catastrophic failure. To start a crack in a composite, on the other hand, you need to apply somewhere between 50 and 70 per cent of its breaking stress. But while it is much more difficult to create the crack in the first place, it will then reach a critical size much more quickly.

This risk of sudden failure means that aircraft using composites are supposed to operate at stress levels far lower than any critical level–even when they experience high loads created by atmospheric turbulence or aircraft wake.

5.2.2 Discussion Questions

Without relying on the text too much and trying to express yourself in your own words, answer the following questions orally. Work with a partner and have each partner take a turn answering the questions.

1. What are composite materials?

2. Why are they now being used in aerospace manufacturing? What are their advantages?

3. What are the pitfalls involved with these materials? What can go wrong?

4. Detection systems need to be established in order to spot when these materials fail. What types of detection devices or procedures could be used? What are the problems associated with conventional procedures such as visual inspections when dealing with composite materials?

5. What parts of an airplane are airplane manufacturers typically using composite materials for?

6. Would you feel comfortable flying in an aircraft built with a significant amount of composite materials? Why or why not?

Fig. 5.2 The Structure of Composite Materials

5.3 Common Chemical Elements from the Periodic Table

There are a number of differences between German and English when it comes to the names of certain chemical elements. Complete the following table with the help of a partner. If you are unsure, check in a dictionary.

Atomic Number	Symbol	German	English
1	H	Wasserstrom	hydrogen
6	C	Kohlenstoff	carbon
7	N	Stickstoff	nitrogen
8	O	Sauerstoff	oxygen
9	F	Flour	fluorine
11	Na	Natrium	sodium
13	Al	Aluminium	aluminium (BE) / aluminum (AE)
14	Si	Silizium	silicon
16	S	Schwefel	sulfur / sulphur
17	Cl	Chlor	chlorine
19	K	Kalium	potassium
22	Ti	Titan	titanium
26	Fe	Eisen	iron (ferro-) as prefix
28	Ni	Nickel	nickel

Atomic Number	Symbol	German	English
29	Cu	Kupfer	copper
30	Zn	Zink	zinc
32	Ge	Germanium	germanium
50	Sn	Zinn	tin
74	W	Wolfram	tungsten
80	Hg	Quecksilber	mercury
82	Pb	Blei	lead
92	U	Uran	uranium

Further Questions

1. *What is meant by atomic number?*

....

....

....

2. *What are the major subatomic particles and where are they found according to standard particle theory?*

....

....

....

....

6 Helicopters

6.1 How Helicopters Work

1.
A helicopter is equipped with one or more power-driven rotors (horizontal propellers) in lieu of fixed wings. It is able to take off and land vertically, to move in any direction, or to remain stationary in the air. The lift developed by a conventional aircraft wing depends on two factors: the angle of attack of the wing (Fig. 6.1) and the velocity of the air in relation to the wing. To obtain the necessary lift, the aircraft must have a forward movement. In the case of a helicopter the (relative) air velocity is produced by the rotation of the rotor blades: when the angle of attack attains a certain value, the lift overcomes the weight of the aircraft. The aircraft then takes off vertically (Fig. 6.2). To achieve horizontal flight, the pilot tilts the rotor forward at a certain angle. This is done by what is known as cyclic pitch change, i.e. changing the pitch of each blade once per revolution. More particularly, the angle of attack of each blade is increased every time it sweeps over the tail of the machine, thereby temporarily developing a greater amount of thrust than the other blades. The thrust developed by the rotor can be resolved into a vertical component (the actual lift that keeps the machine in the air) and a forward component (which propels the machine horizontally) (Fig. 6.3). Each blade can swivel about its longitudinal axis and its pitch is changed cyclically, through a linkage system, by a so-called swash-plate, which performs a sort of wobbling rotary motion around the shaft and swivels the blades to and fro as they rotate. The tilt of the swash plat can be varied by the pilot, and the tilt of the rotor follows the tilt of the plate.

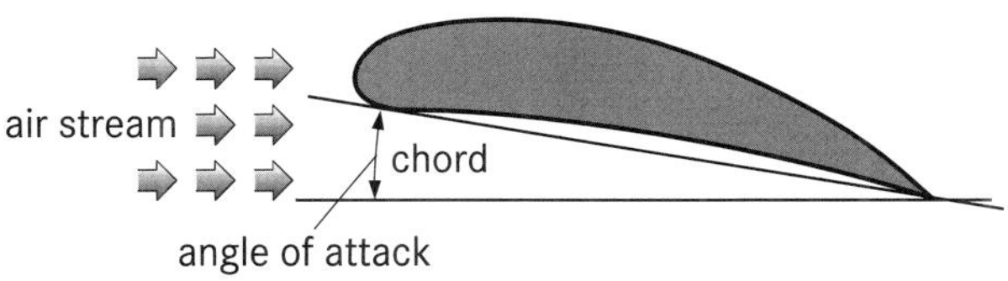

Fig. 6.1 Angle of Attack

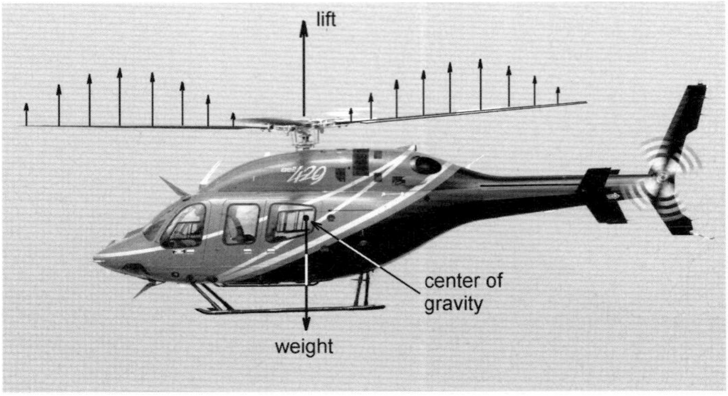

Fig. 6.2 Vertical Take-Off

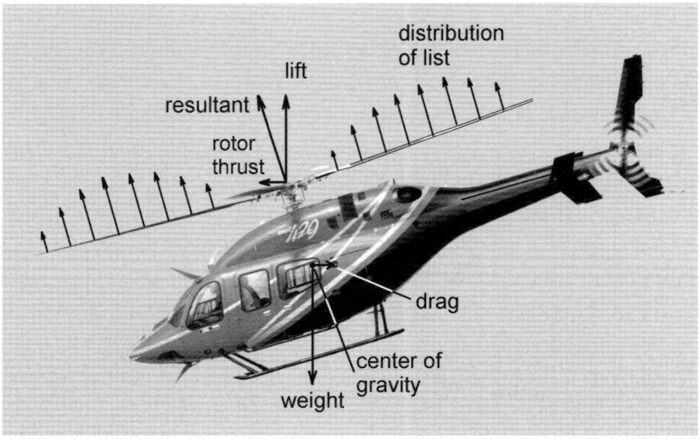

Fig. 6.3 Forward Flight

2.
The blade root hinges (called lag hinges) shown in Fig. 6.4 allow "blade flapping", as represented in Fig. 6.5. if there were no hinges, tilting of the plane of rotation of the rotor blades relative to the helicopter would cause a periodic change in the speed of the blades. This would produce severe stresses in the blades; these stresses are relieved and cancelled by the hinge. Motion about the hinge enables the blade to rotate at constant speed irrespective of how much the rotor is tilted. In forward movement of the helicopter, the velocity due to blade rotation and that due to forward speed are added together, i.e. they intensify each other, on the advancing side of the rotor; on the retreating side, however, they are subtracted from each other. If the rotor blades were rigidly fixed to the shaft, the lift would vary cyclically and cause the helicopter to roll. This is prevented by the hinge. Instead, the blade flaps cyclically as it rotates.

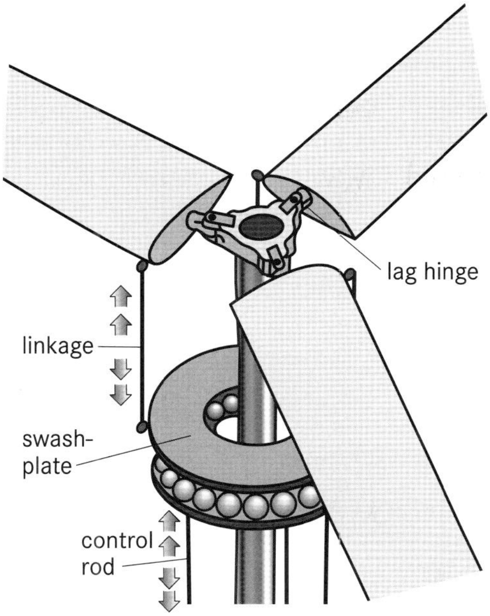

Fig. 6.4 Rotor Head

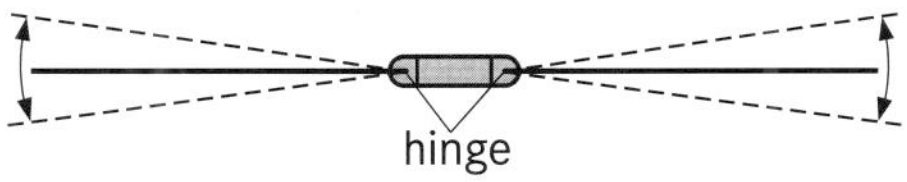

Fig. 6.5 Blade Flapping

3.
The rotation of the rotor tends to cause the fuselage of the aircraft to rotate in the opposite direction (on the principle that any action calls forth a reaction). To prevent this, the single-rotor helicopter is provided at its tail with a small propeller producing a counteracting sideways thrust. Alternatively, the helicopter may have two rotors which revolve in opposite directions and thus counterbalance each other (Fig. 6.6).

(Source: How things work)

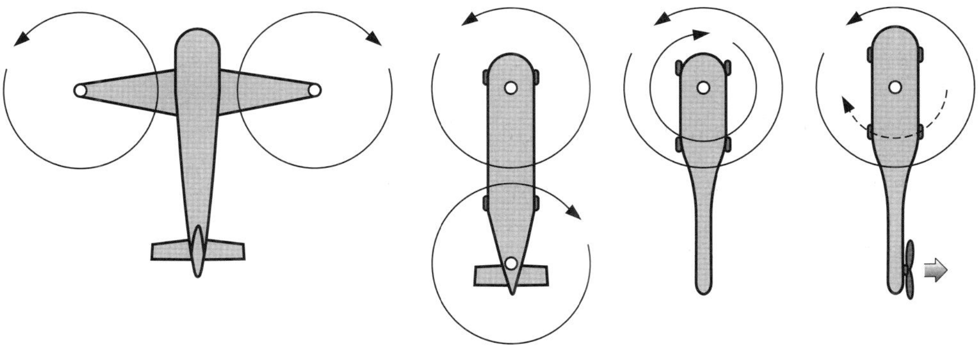

Fig. 6.6 Methods of Preventing Fuselage Rotation

6.1.1 Exercise A: Comprehension Questions

Based on the preceding text, answer the following questions using full sentences.

1) At what point does a helicopter take off?

The rotor blades are rotating and when the angle of attack attains a certain value, the lift ~~be~~ is higher than the weight of the helicopter which leads to a vertical take-off.

2) How does a helicopter pilot achieve horizontal flight?

By tilting the rotor forward at a certain angle. This is achieved by changing the pitch of each blade once per revolution.

3) What is the function of swash plates?

They are used to swivel the blades to and fro along their longitudinal axis as they rotate.

denn: man muss blade-Geschw. & horizon. Geschw. addieren ← because of periodic change in the speed

4) What is "blade flapping"? → it lightens the stress in the blades.

The blades are attached to the helicopter by hinges that allow the blades a certain up and down movement called "blade flapping".

5) What additional feature does a single-rotor helicopter need?

It needs a small propeller at its tail in order to prevent the fuselage from rotating in the opposite direction than the rotor.

Fig. 6.7 Control Panels on a Helicopter

6.1.2 Exercise B: Vocabulary Questions

Answer the following vocabulary questions based on the terms found in the text.

Taken from paragraph one:

a) What is another way of saying *in lieu of*?

instead of

b) What is the difference between *attain* and *obtain*?

c) Create a sentence using *thereby*?

Take from paragraph two:

d) What is the meaning of *plane* in this context?

e) How would you translate *hinges*?

f) What is another way of saying *irrespective of*?

= regardless of ...
= no matter ...

Fig. 6.9 The Interior of a Helicopter Cabin

Fig. 6.8 A Helicopter in Flight

6.1.3 Exercise C: Words of Motion

The following words of motion were mentioned in the text.

revolution	sweep	roll	to and fro
rotation	propel	flap	tilt
swivel	wobbling		

1) Which words involve circular motion?

2) Which ones necessarily imply 360 degrees?

3) Which ones involve repetitive back and forth or up and down motion?

4) Which ones do not involve a fixed axis?

5) Which imply speed?

Fig. 6.10 Eurocopter AS-365 Dauphin

6.1.4 Exercise D: Translation

Translate the following text from German into English.

Um aus dem Schwebeflug in den senkrechten Steigflug überzugehen, muss der Einstellwinkel aller Rotorblätter gleichmäßig durch den kollektiven Blattverstellhebel (Pitch) vergrößert werden. Eine Vergrößerung des Rotorschubes und des Auftriebes bringt auch eine Vergrößerung des Blattwiderstandes mit sich, der durch erhöhte Triebwerksleistung überwunden werden muss. Das Angleichen der Motorleistung kann sowohl durch einen Drehgasgriff als auch automatisch über Triebwerksleistungsregler erfolgen. Sobald der Hubschrauber in einen gleichmäßigen Steigflug übergegangen ist, kann der Rotorschub auf eine Größe reduziert werden, die dem Gewicht plus dem Widerstand der Zelle entspricht.

(Source: www.hubschraubermuseum.de)

6.2 Describing the Movement of a Graph

When describing graphs in English it is important to be able to express yourself in a varied manner, so that you do not always repeat the same expressions. Otherwise, your description becomes rather dull quickly. Furthermore, you should try to avoid expressions such as “it goes up a little” or “the graph goes down a lot”, which are seen to be somewhat simplistic descriptors and would only serve to discredit your expertise as an aerospace engineer.

The following five movements of a graph will be examined more closely in this section:

a) upward movement

b) downward movement

c) no movement

d) peaks

e) troughs

Upward Movement

The following verbs are used to express the upward movement of a graph:

- ~~the graph goes up (very basic form~~)
- the graph increases
- the graph rises
- the graph rockets up or booms kein „steeply", ... danach (schon dramatisch genug)

Important Reminder

Rise versus Raise

Do not confuse the verb *raise* with the verb *rise*. *Rise* is used only when there is no direct object. *Compare:*

“The wind velocity has risen.”

with

“The decrease in pressure has raised the velocity of the airflow over the airfoil”.

In the second sentence, there is a direct object. What has been raised? The velocity.

Rise is an intransitive verb (used without a direct object) and *raise* is a transitive verb (used with a direct object). This phenomenon is similar in German with *legen* und *liegen*. Which one is transitive and which one is intransitive, do you think?

Downward Movement

The following verbs are used to express the downward movement of a graph:

- the graph goes down (very basic form)
- the graph decreases
- the graph falls
- the graph declines
- the graph drops
- the graph plunges or plummets

Important Point

Fall versus Fall Down

When describing a graph, do not use the verbal expression *fall down*, which is reserved for other contexts, such as *to fall down the stairs or to fall down a hill.* For graphs, we use simply the verb *fall.*

No Movement

The following verbs are used to describe a portion of a graph where there is no movement:

- the graph stays the same
- the graph remains constant
- there is no change
- the graph stabilizes
- the graph levels off / tapers off (indicates change that is followed by very little or no movement)

Peaks

The following verbs are used to describe a graph where a maximum value is reached (either global or local).

- the graph peaks at (x value)
- the graph reaches a peak at (x value)
- the graph reaches its maximum value at (x)

Troughs

The following verbs are used to describe a graph where a minimum value is reached (either global or local).

- the graph bottoms out at (x value)
- the graph reaches its minimum value at (x value)
- the graph reaches its lowest point at (x value)

Important Point: Verb Tenses

When describing a graph, it is uncommon to use the progressive or continuous forms (i.e. those with -ing), as they tend to describe an action in progress either in the present, past or future. Instead, we tend to use the simple forms in English, i.e. present simple, past simple or future simple. The following examples that could be used to describe a graph illustrate this point:

The cost of manufacturing a helicopter increased between 1950 and 1980.

The ductility of steel falls dramatically after the steel specimen has been stretched beyond a certain point.

The efficiency of aircraft engines will increase only slightly in the coming years.

Amount/Time

Should you wish to describe the amount of change that has taken place in the graph you are describing or the time in which it this change has occurred, the following adverbs are particularly useful:

Amount

- hardly/barely (placed before the verb)
- negligibly
- slightly
- somewhat
- noticeably
- considerably
- significantly

Time

- slowly
- gradually
- suddenly
- quickly/fast/rapidly/swiftly

Amount or Time

- steeply
- dramatically
- sharply

Further Helpful Expressions

The following are some further helpful expressions when describing a graph:

- the graph fluctuates (goes up and down quickly over a period of time)
- the graph slumps (drops significantly)

6.2.1 Exercises

Draw a graph for each of the following descriptions.

1) Plot temperature against pressure. The line on the graph begins at the origin, increases linearly and dramatically, reaches a peak half way through the x axis and then declines gradually.

2) The next graph depicts one-off development costs versus market share. The slope begins at the middle of the y axis and then fluctuates considerably before leveling off at a higher market share.

3) In this graph, the lift-drag ratio is plotted as a function of Mach number. The ratio is relatively high at the beginning and then slumps after breaking the sound barrier. At approximately Mach 2, the lift-drag ration bottoms out and more or less levels off at higher supersonic speeds.

6.3 Agreeing and Disagreeing

Place the following expressions relating to agreeing and disagreeing in the correct category in the subsequent table. Feel free to add to the table any expressions you might know of that are not listed below.

I agree (with you) up to a point, but ...

I couldn't agree more.

I think so, too.

I'm not sure I agree with you.

That argument just doesn't hold.

I beg to differ with you.

I'm of two minds about that.

I don't agree with that at all.

I'm of the same opinion.

To a certain extent, I agree with you. However, ...

I'm not sure where I stand on that.

Expressing agreement
Saying you partly agree
Disagreeing in a less direct way
Disagreeing in a more direct way
Saying you are undecided

6.3.1 Text: Helicopter with Flapping Rotors

a) *Read the text "In a Flap" below and decide whether you think this design is a viable one. Why or why not?*

Traditional helicopters have a vertical tail rotor to stop the main rotor from spinning the craft endlessly round and round. But running that second rotor wastes around eight per cent of the chopper's fuel, says Vladimir Savov, an expert in aerodynamics. So to save that fuel, Savov has invented the "rotopter", a helicopter that flaps its rotors.

Helicopters without tail rotors use counter-rotating main rotors or elaborate fans on the tail boom, but Savov says his version will be far cheaper. To fight the tendency of a helicopter to turn in the opposite direction to its spinning rotor system, Savov has removed the engine that spins the main rotors. Instead, flapping the rotors up and down makes them spin on their own.

Savov's design involves an engine driving a crank that creates an up-and-down reciprocating motion that in turn drives two separate rotor arms. By flapping the rotors up and down, and carefully choosing the angle each rotor's aerofoil presents to the air during its cycle, the rotors begin to rotate on their own.

"There is no rotating moment transmitted to the rotopter rotor blades, so there is no torque reaction and so no need for a tail rotor,"

(Source: "In a Flap", 24 June 2000 in New Scientists by Paul Marks (adapted and abridged))

Fig. 6.11 Eurocopter 135 with fly-by-wire/fly-by-light system

b) *Discuss this new design with a partner using the structures you learned above relating to agreeing and disagreeing. Try to reach a consensus.*

6.3.2 Writing

Your boss has asked you to assess the viability of this new design. Write a short e-mail to your boss and try to get across your point of view in a persuasive manner.

7 The Future of NASA

7.1 NASA's Vision

Fig. 7.1 ISS

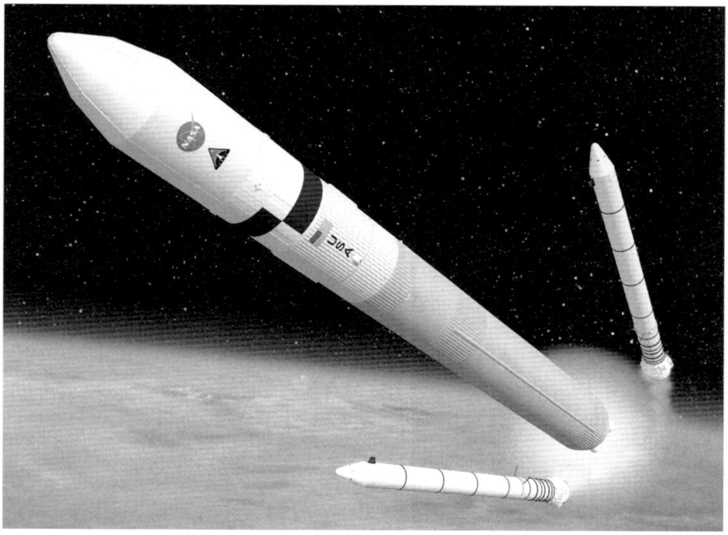

Fig. 7.2 Ares V

Fig. 7.3 ExoMars Rover

The following is a summary of NASA's latest vision and priorities for the future:

- extending the life of the International Space Station, likely to 2020 or beyond
- developing new launch and space transportation technologies, nimble construction capabilities on orbit, and new operations capabilities
- sending humans back to the Moon and on to Mars and asteroids, all with the collaboration of other countries
- supporting the commercial spaceflight industry to make travel to low Earth orbit and beyond more accessible and more affordable, thereby creating the opportunity for thousands of people to visit and possibly even to live in low Earth orbit
- investing in the scientific observation of our planet, especially with regard to climate change

(Source: NASA-2011_Budget Administrator Remarks)

Discussion Questions

In groups of two or three, discuss the following questions.

A) What are the merits of this vision? Do you agree with its focus? Why or why not? Are there any other priorities which NASA should set, e.g. more pure scientific research, more money for the Hubble Telescope? What priorities would you set for NASA?

B) Why has the shuttle program not been an unmitigated success? Are fundamental design principles and/or features to blame? What caused the Columbia accident?

C) What options are open to NASA for launching payload and crew into space once the shuttle fleet retires?

7.2 The Shuttle Program

The Space Shuttle is the world's first reusable spacecraft, and the first spacecraft in history that can carry large satellites both to and from orbit. The Shuttle launches like a rocket, maneuvers in Earth's orbit like a spacecraft and lands like an airplane. Each of the three Space Shuttle orbiters - Discovery, Atlantis and Endeavour - were designed to fly at least 100 missions.

Columbia was the first Space Shuttle orbiter to be delivered to NASA's Kennedy Space Center, FL, in March 1979. Columbia and the STS-107 crew were lost Feb. 1, 2003, during re-entry. The Orbiter **Challenger** was delivered to KSC in July 1982 and was destroyed in an explosion during ascent in January 1986. **Discovery** was delivered in November 1983. **Atlantis** was delivered in April 1985. **Endeavour** was built as a replacement following the Challenger accident and was delivered to Florida in May 1991. An early Space Shuttle Orbiter, the **Enterprise**, never flew in space but was used for approach and landing tests at the Dryden Flight Research Center and several launch pad studies in the late 1970s.

The Space Shuttle consists of three major components (Fig. 7.4): the Orbiter which houses the crew; a large External Tank that holds fuel for the main engines; and two Solid Rocket Boosters which provide most of the Shuttle's lift during the first two minutes of flight. All of the components are reused except for the external fuel tank, which burns up in the atmosphere after each launch.

Fig. 7.4 The components of the Space Shuttle system: orbiter, external tank and solid rocket boosters

Length: Space Shuttle: 184 feet; Orbiter: 122 feet
Height: Orbiter on runway: 57 feet
Wingspan: 78 feet
Lift-off Weight: 4.5 million pounds
Orbit: 115 to 400 statute miles
Velocity: 17,321 mph

The longest the Shuttle has stayed in orbit on any single mission is 17.5 days on mission STS-80 in November 1996. Normally, missions may be planned for anywhere from five to 16 days in duration. The smallest crew ever to fly on the Shuttle numbered two people on the first few missions. The largest crew numbered eight people. Normally, crews may range in size from five to seven people. The Shuttle is designed to reach orbits ranging from about 185 kilometers to 643 kilometers (115 statute miles to 400 statute miles) high.

The Shuttle has the most reliable launch record of any rocket now in operation. Since 1981, it has boosted more than 1.36 million kilograms (3 million pounds) of cargo into orbit. More than 600 crew members have flown on its missions. Although it has been in operation for almost 30 years, the Shuttle has continually evolved and is significantly different today than when it first was launched. NASA has made literally thousands of major and minor modifications to the original design that have made it safer, more reliable and more capable today than ever before.

Fig. 7.5 Space Shuttle, mobile launcher platform and rotating service structure

Since 1992 alone, NASA has made engine and system improvements that are estimated to have tripled the safety of flying the Space Shuttle, and the number of problems experienced while a Space Shuttle is in flight has decreased by 70 percent. During the same period, the cost of operating the Shuttle has decreased by one and a quarter billion dollars annually - a reduction of more than 40 percent. At the same time, because of weight reductions and other improvements, the cargo the Shuttle can carry has increased by 7.3 metric tons (8 tons.)

In managing and operating the Space Shuttle, NASA holds the safety of the crew as its highest priority.

(Source: NASA)

7.3 Delayed Discovery Launch

7.3.1 Vocabulary Exercise

The following is a list of difficult words or expressions taken from an audiotext from www.npr.org relating to one of the Discovery launches. *Match the words/expressions with their definitions.*

1) chunk
2) lethal
3) foam
4) debris
5) liberated
6) condensation
7) validate
8) bellows
9) deemed
10) the clock is ticking
11) requisite
12) wrinkle
13) as it happens

a) .... expandable component used to protect screws, shafts, and other components from liquids, dirt and other debris

b) .... considered, regarded

c) .... sufficient to cause death

d) .... scattered fragments of something destroyed or wrecked

e) .... in fact, in reality

f) .... time is running out

g) .... a frothy, bubbly substance in a cellular mass

h) .... released

i) .... the reduction of a gas or solid to a liquid

j) .... needed by circumstances

k) .... a slight crease in a flexible surface

l) .... a thick solid slice or piece of something firm or hard

m) .... confirm

7.3.2 Audiotext: Delayed Discovery Launch

"Fuel Tank Issue Delays NASA Shuttle Launch", April 29, 2005

http://www.npr.org/templates/story/story.php?storyId=4625287

Listen to the radio broadcast from National Public Radio relating to the one of the Discovery launches and answer the following questions. Point form instead of full sentences is sufficient. Enter the web address listed above into your Internet browser in order to gain access to the radio broadcast.

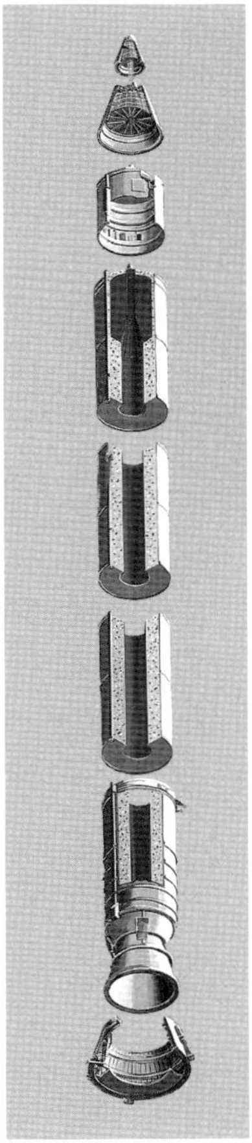

Fig. 7.6 Space Shutttle propulsion systems

1) Why has the latest Discovery launch been delayed?

....

....

....

....

2) Why did Columbia crash?

....

....

....

....

3) What is the latest potential danger?

....

....

....

....

4) Why is it a problem? How does this situation arise?

....

....

....

....

5) How has NASA been working to solve this problem?

....

....

....

....

6) What are the shuttles currently being used for?

7) What is another use that has been proposed?

Fig. 7.7 Space Shuttle launch

7.4 Ares Program

7.4.1 Introductory Text

Ares I is an in-line, two-stage rocket topped by the Orion capsule, its service module and a launch abort system. The combination of the rocket's configuration and Orion's launch abort system, which can move astronauts away quickly in case of a launch emergency, will improve crew safety.

The launch vehicle's first stage is a single, five-segment reusable solid rocket booster, derived from the Space Shuttle Program's four-segment reusable solid rocket booster, which burns a specially formulated and shaped solid propellant called polybutadiene acrylonitrile (PBAN). A newly designed forward adapter called a frustum will mate the vehicle's first stage to the second, and will be equipped with booster separation motors to disconnect the stages during ascent.

The second, or upper, stage is being designed at Marshall Space Flight Center. Much like the upper stage for the Ares V cargo launch vehicle, the Ares I upper stage is propelled by a J-2X main engine fueled with liquid oxygen and liquid hydrogen.

Fig. 7.8 Orion in orbit

The J-2X is an evolved variation of two historic predecessors: the powerful J-2 upper-stage engine that propelled the Apollo-era Saturn IB and Saturn V rockets to the Moon and the J-2S, a simplified version of the J-2 developed and tested in the early 1970s.

Ares I has two missions: lofting four astronauts (or cargo) to the International Space Station–approximately 52,000 pounds–or to low-Earth orbit for rendezvous with the Ares V Earth departure stage for missions to the Moon (56,000 pounds).

During the first two-and-a-half minutes of flight, the first stage booster powers the vehicle to an altitude of about 187,400 feet (35.5 miles) and a speed of Mach 5.7. After its propellant is spent, the reusable booster separates and the upper stage's J-2X engine ignites and powers the Orion spacecraft to an altitude of about 425,328 feet (80 miles). Then, the upper stage separates and Orion's service module propulsion system completes the trip to a circular orbit of 976,800 feet (185 miles) above Earth.

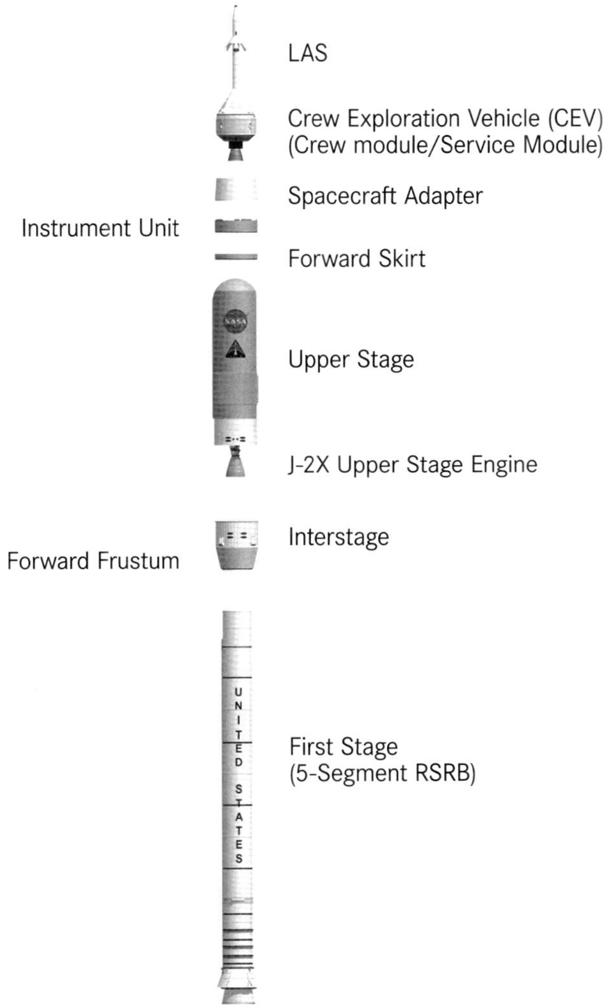

Fig. 7.9 Ares I in-line, two-stage rocket configuration topped by the Orion crew vehicle

Once in orbit, the Orion and its service module will rendezvous and dock either with the space station or with the Altair lunar lander and Earth departure stage that will send the astronauts on their way to the moon.

(Source: NASA)

7.4.2 After the Shuttle (Summary)

In groups of two (maximum three), create a table that summarizes how the Ares I differs from the Space Shuttle. Focus on the purpose, design features, function and any other characteristics you feel are important. Be sure to organize your information in a table such that someone who has not read the article can easily understand the content.

Criterion	Space Shuttle	Orion Spaceship
Purpose/Function		
Design / Features		
Capacity		
Costs		
Additional Features		

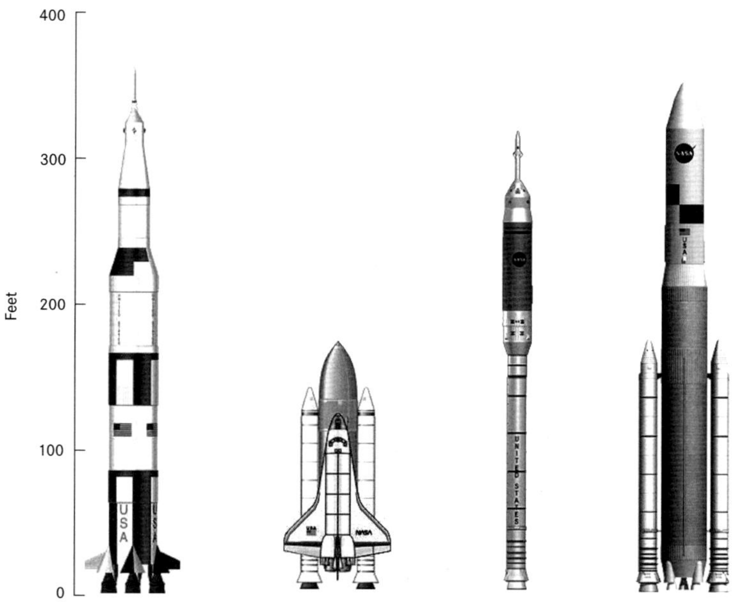

Fig. 7.10 Comparison of the Saturn V, Space Shuttle, Ares I and Ares V

7.5 False Friends

A false friend is typically a word in one language, in this case English, which resembles a word in another language, in this case German, whose meaning, however, differs from the other word.

For example, **become (= werden)** and **bekommen (= get, receive)** are false friends, as are **gift (= Geschenk)** and **Gift (= poison).**

The following pairs of words are false friends. Provide the correct translation for both words in the pair and then write a sentence showing the true meaning of the English word.

a)	isolated	Isoliert (as in *isoliertes Kabel*)
b)	construction	Konstruktion
c)	control	kontrollieren
d)	technique	Technik (as in *in neue Technik investieren*)
e)	technique	Technik (as in *Elektrotechnik*)
f)	beamer	Beamer
g)	eventually	eventuell
h)	actually	aktuell
i)	special	speziell
j)	sensible	sensibel
k)	prove	prüfen
l)	promotion	Promotion
m)	consequently	consequent
n)	prospects	Prospekte
o)	defect	defekt
p)	delicate	delikat
q)	floor	Flur
r)	edge	Ecke

8 Space Elevators

8.1 Satellites on a Leash

The children's story „Jack and the Beanstalk“ could soon move from the **realm** of a fairy tale to practical reality, if work on the project of a space elevator, now at an advanced engineering stage, receives the go-ahead for construction.

The idea behind the project is that of a space elevator, built with a geosynchronous satellite as the base from which to build the elevator. By using a **counterweight**, a cable would be lowered from geosynchronous orbit – at an altitude of about 37,000 km – towards Earth's surface while the counterweight would be extended from the satellite in the opposite direction, away from the Earth. The competing forces of gravity at the lower end and outward centrifugal acceleration at the far end would keep the cable under tension and stationary over a single position on Earth. The total cable length would be about 100,000 km, where a counterweight of only about 650 t is required.

Fig. 8.1 An illustration of how a space elevator might appear

This cable could **ascend** by mechanical means to an orbit around the Earth. To place a spacecraft in geosynchronous orbit, a climber, or electric lift, would simply ascend to that altitude and release the payload. To place a spacecraft in any other orbit would require a small engine to **attain** the proper orbital velocity. At the far end of the cable the spacecraft would have sufficient energy to escape from Earth's gravity by simply **detaching** itself from the cable in order to travel to another planet, for example. There would, therefore, be huge advantages in using such a cable in lieu of a rocket to place spacecraft into orbit or to launch them into space. **Proponents** of the space elevator suggest it could lower the cost of placing a satellite into space from US$20,000/kg to US$100/kg.

The principal scientist working on the project, Dr Bradley C Edwards, thinks that most of the system's primary questions, including safety and environmental **hazards**, design and outline of the whole system, have been answered. He and his team are now focusing on the system's engineering and dynamics. Edwards hopes construction of the elevator will start in about five years. He said one of the most **daunting** technical problems associated with the concept has been the tension on the cable.

Tethering the Sky

The cable would need a strength of 100 GPa, many times that of steel, and also greater than that of kevlar, carbon fibre, quartz, diamond, or spider silk. No known material had the necessary **tensile strength** until the discovery of **buckyballs**, **buckytubes** and similar structures. When buckytubes, also called carbon nanotubes, and similar structures were first discovered, it was quickly noted that they possessed an incredibly high tensile strength. However, it is doubtful anybody would have, at the time, predicted the use of those materials in a cable to support a space tower.

The discovery of those carbon nanotubes was one of the main factors responsible for the **revival** of the space elevator concept. Late last year, the Los Alamos National Laboratory, in New Mexico, US, co-sponsored a conference on the space elevator. The **keynote address** was given by Arthur C Clarke, of science fiction fame, speaking via satellite from his home in Sri Lanka. He has written a novel ("The Fountains of Paradise") around a space elevator. The original idea is more than a century old, however. In the late 19th Century, the Russian space scientist and visionary, Konstantin E Tsiolkovsky, proposed a tower high up in the sky attached to a "**celestial** castle" in orbit around Earth, with the centrifugal force of the castle holding up the tower.

The concept was taken up by Edwards, then at Los Alamos, who sent a **proposal** to NASA's Institute for Advanced Concepts. The Institute provided him with a US$570,000 **grant** to work on the concept. Under that grant Edwards, now director of research with the Institute for Scientific Research in Fairmont, West Virginia, and his team carried out an investigation in two phases.

(Source: Engineering World, April/Mai 2004 Issue)

8.1.1 Vocabulary

Match the English words taken from the text above and listed on the left in the table below with their German counterparts on the right. The words in English have also been marked in bold in the text.

1. ascend	a. Zugfestigkeit
2. attain	b. Gegengewicht
3. buckyball, buckytube	c. Himmels-
4. celestial	d. Bereich, Gebiet
5. counterweight	e. Gefahr
6. daunting	f. erreichen
7. detach	g. aufsteigen
8. grant (n)	h. Förderung
9. hazard	i. gewaltig, entmutigend
10. keynote address	j. Befürworter
11. proponent	k. Wiedererwachen
12. proposal	l. anbinden
13. realm	m. abtrennen
14. revival	n. Grundsatzrede
15. tensile strength	o. Vorschlag, Antrag
16. tether	p. netzartige Strukturen der Nanotechnik

8.1.2 Comprehension Questions

After reading the text on space elevators on the preceding pages, answer the following questions.

1) How would you describe what a space elevator is?

....

....

....

....

2) Describe briefly how it works.

....

....

....

....

3) What is one important advantage of such a machine?

4) What has been one of the difficulties with the engineering side of it?

5) What solution is being examined to address this difficulty?

6) Sketch how you imagine this machine.

8.2 Summarizing/Paraphrasing/Clarifying

The text "Satellites on a Leash" continues and is divided into two further sections. Working in pairs, choose one of the texts below and summarize it orally for your partner. (Refer back to 4.5 for language structures related to paraphrasing.)

8.2.1 Text One

Beam me up, Scotty

Phase I laid out a detailed description of a possible space elevator program. A small, carbon nanotube composite cable capable of supporting 495 kg payloads would be deployed from a geosynchronous orbit using shuttles and liquid or solid fuel upper stages. Climbers would be sent up the initial cable adding cables to the first to increase its strength. They would travel in an apparatus that would lift itself upward with a couple of treads squeezing lightly onto the ribbon. After about two years, a cable capable of supporting payloads up to 20,000 kg would be complete. The power for the climbers would be beamed up in a free-electron laser and received by photocells.

The spent initial spacecraft and climbers would become counterweights at the space end of the 100,000 km-long cable. An ocean-going platform would be used for the Earth anchor. This anchor would be mobile and able to move the cable out of the way of low-orbit satellites. The anchor location would be in the Pacific Ocean, about 1500 km west of the Galapagos Islands to avoid lightning, hurricanes, strong winds, and clouds.

The specific cable design would be a curved and tapered ribbon with a width increasing from Earth to the geosynchronous orbit and back down to the far end. Deviations in the cable's cross-sectional dimensions would be implemented to reduce the risk of damage from meteors and wind. During Phase II the team concentrated on answering any remaining basic feasibility questions.

(Source: Engineering World, April/Mai 2004)

Fig. 8.2 ESA‘s wind-monitoring satellite Aeolus

8.2.2 Text Two

The primary areas studied under Phase II included: large-sacle carbon nanotube production, cable production and design, the power beaming system, the effects of the weather at the anchor site, anchor design, the structure's environmental impact, placing payloads in Earth's orbit, elevators on other planets, possible tests of the system, major design tradeoffs, budget estimates, and a independent review of the program.

Long Lines from Tiny Tubes

Edwards said the most critical element in the development of the space elevator is the design, construction and testing of the carbon nanotube ribbon segments. He said it is absolutely critical to have ultra-high-strength material (100 GPa) in a usable form. He and his team have been testing carbon nanotubes from several suppliers.

He said the team is looking into the development of a ribbon with individual small fibres, rather than a sheet of carbon nanotube composites. He said this change was partly due to an improved understanding of the degradation of the ribbon and partly due to a better understanding of the production of carbon nanotube composites. The team is now developing a technique for producing ultra-strong individual fibres about 10 μm in diameter and lengths ranging from several meters to kilometers.

(Source: Engineering World, April/Mai 2004)

8.3 Attending a Conference

8.3.1 Exercise A

Read the information below relating to an upcoming conference on Space Science and Engineering. Identify

- *the key information*
- *the information that is missing if you were planning on giving a presentation at the conference*

September 27 – 30, 2010

The 10th Australian Space Science Conference (ASSC), is to be held in Brisbane at the University of Queensland, Brisbane. This will be the fourth ASSC jointly sponsored and organised by the National Committee for Space Science (NCSS) and the National Space Society of Australia (NSSA). The ASSC is intended to be the primary annual meeting for Australian research relating to space science. It welcomes space scientists, engineers, educators, and workers in Industry and Government.

This year's ASSC will run in conjunction with the NCSS's workshop on implementing Australia's first Decadal Plan for Space Science, which is ready to be published. This one-day workshop will discuss the Plan and Government's responses, better link the scientific community and associated stakeholders in Government and industry, and start implementing the Plan's recommendations.

The scope of the conference covers fundamental and applied research that applies to space technologies, and includes the following:

- Space science, including space and atmospheric physics, Earth observation and remote sensing from space, planetary sciences, astrobiology and life sciences, and space-based astronomy and astrophysics.

- Space engineering, including communications, navigation, space operations, propulsion and spacecraft design, testing, and implementation.
- Space industry
- Government, international relations and law
- Education and outreach

This years conference will feature streams sponsored by the Australian Space Research Institute (ASRI).

There will be a special session on successful, submitted, and future applications for Australian Space Research Program funding.

Key Dates

- 5 April - Opening date for Abstracts
- 11 July - Closing date for Abstracts (Presentation or Poster)
- 11 July - Conference Registration Opens
- 25 July - Notification (Acceptance/Rejection of Abtract)
- 2 August - Draft Program is available
- 23 August - Early-Bird Registrations Close
- 24 August - Standard Conference Registrations Open
- 28 August - Final Program is available
- 27–30 September - Conference Registration (Late)
- 2 November - Deadline for Final Papers

Researchers and students are invited to submit an abstract for either a paper or poster. Full details are contained in the documents which can be downloaded from our website.

[Adapted from 10th Australian Space Science Conference (ASSC); National Committee for Space Science and National Space Society of Australia]

8.3.2 Exercise B

*You are interested in attending this conference. Working with a partner who will act as your direct superior at work, role play a conversation you would have with him or her in which you try to convince him or her to allow you to attend the conference on behalf of your department. You are only prepared to attend, however, if your superior decides to reimburse you for all costs incurred (*alle anfallenden Kosten erstatten*).*

Helpful Expressions

- I was wondering if it would be possible to...
- Given the importance of...
- Do you see any way to...
- As you are (already/likely) aware,...
- It's essential/crucial/of utmost importance...

8.4 Uncountable Nouns

One of the concerns with space elevators is that the interference caused by the weather would jeopardize the operation and safety of the apparatus. The word "interference" is an uncountable noun; that is, you cannot say in English "one interference" or "two interferences" and so on and so forth. Therefore, an indefinite article (*a/an*) cannot be used before the noun "interference". Another uncountable noun in English is "equipment".

Make a list of other uncountable nouns in English.

....

....

....

....

....

Dealing with Singular and Plural Quantities

As mentioned earlier, you cannot use "a, an, one, two, etc." in conjunction with uncountable nouns. Therefore, English employs other strategies when it comes to expressing the singular or plural form of uncountable nouns.

a piece of

a bit of / a lot of

some

much/little (Note that "much" is mostly used in negative sentences, e.g. There wasn't much atmospheric interference with the tests we conducted.)

e. g. *The cost of acquiring this piece of equipment exceeds our budget for the entire year.*

9

Electrical Rocket Propulsion

9.1 Introduction to Electrical Rocket Propulsion

Overview

Electrical and electromagnetic rockets differ fundamentally from chemical rockets with respect to their performance limitations. Chemical rockets are essentially **energy-limited**, since the quantity of energy (per unit mass of propellant) which can be released during combustion is limited by the fundamental chemical behavior of propellant materials. If, on the other hand, a separate energy source (e.g. nuclear or solar) is used, much higher propellant energy is possible. Further, if the temperature limitations of solid walls could be made unimportant by direct electrostatic or electromagnetic propellant acceleration without necessarily raising the fluid and solid temperatures, there would be no necessary upper limit to the energy which could be added to the propellant. However, the rate of conversion of nuclear or solar energy to electrical energy and thence to propellant kinetic energy is limited by the mass of conversion equipment required. Since this mass is likely to be a large portion of the total mass of the vehicle, the electrical rocket is essentially **power-limited**.

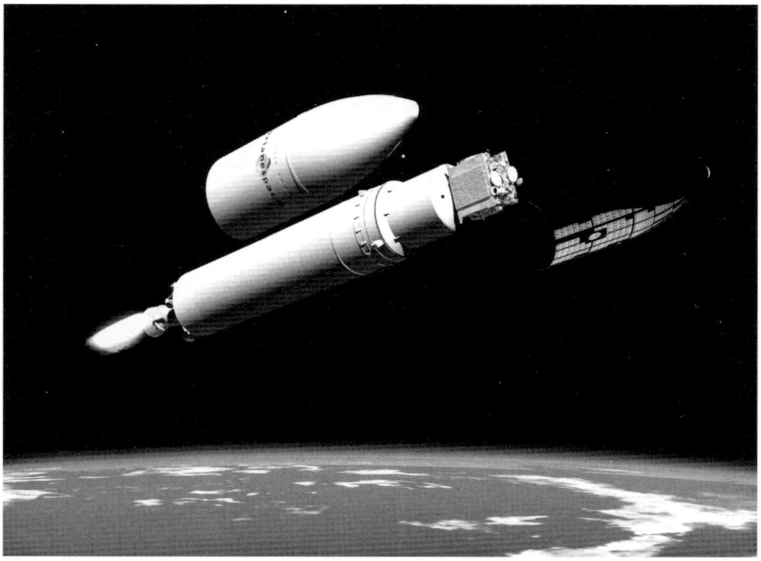

Fig. 9.1 Ariane 5 releasing its fairing

Electrostatic and electromagnetic rockets convert electrical energy directly to propellant kinetic energy without necessarily raising the temperature of the working fluid. For this reason the specific impulse is not limited by the temperature limitations of wall materials, and it is possible to achieve very high exhaust velocities, although at the cost of high power consumption.

Types

In general, then, electrical rockets should be able to develop considerably higher specific impulse than chemical or nuclear ones. It has been shown, however, that this gain in specific impulse requires massive energy-conversion equipment, so that electrical rockets generally have low thrust, perhaps only one-thousandth of the vehicle weight in the earth's gravitational field. For this reason they are mainly restricted to space missions during which gravitational forces are very nearly balanced by inertial forces. Low accelerations are quite acceptable, since the journeys are of long duration.

The propellant of an electrical rocket consists of either discrete charged particles which are accelerated by electrostatic forces, or a stream of electrically conducting fluid which is accelerated by electromagnetic or pressure forces (or both).

Electrostatic rockets are subdivided into categories depending on the nature of their propellant; charged single atoms or ions, and charged, relatively massive, multi-atom particles called colloids. Ion rockets are further subdivided according to the process by which the ions are formed. Bombardment ionization results from the collision of a high-speed electron with a single propellant atom in the gaseous phase. Contact ionization is an intersection between a single propellant atom and a suitable hot surface.

(Source: "Mechanics and Thermodynamics or Propulsion" by Hill and Peterson, 3rd printing, 1965. Reprinted by permission of Pearson Education, Inc., Upper Saddle River, New Jersey)

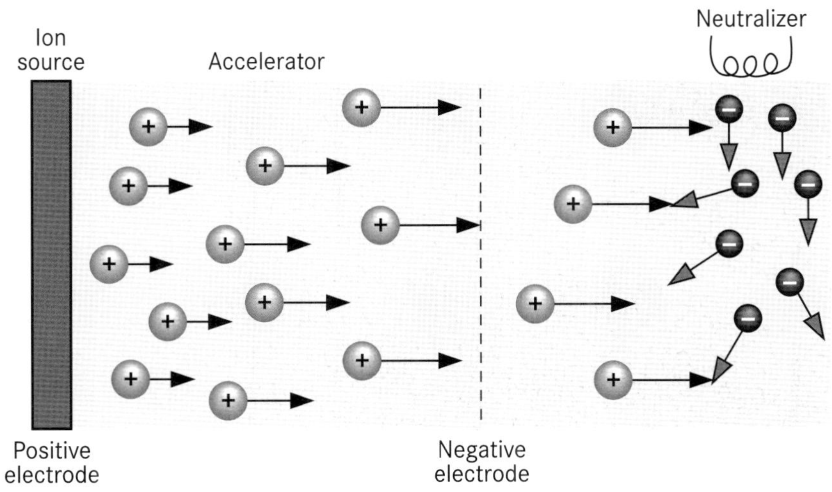

Fig. 9.2 Schematic representation of the functioning of an electromagnetic accelerator

The arcjet, or plasmajet as it is sometimes called, is not really an electrical acceleration device. Electrical power is used simply to heat the propellant to a very high temperature, after which it is accelerated in a conventional nozzle. The fact that the propellant is electrically conductive is incidental to the acceleration process. Arcjets are therefore electrothermal in nature.

Other types of electrothermal thrusters include resistojets and RF-heated thrusters. The resistojet essentially uses a filament to heat a propellant gas (not plasma). The RF-heated thruster uses radio-frequency waves to heat a plasma in a chamber and potentially could reach somewhat higher exhaust velocities.

Electromagnetic accelerators utilize the conductivity of the plasma propellant to create an electromagnetic accelerating force acting as a 'body force' within the plasma.[1] There are three principal methods of accelerating plasmas electromagnetically; these are analogous to the behavior of shunt, series and induction motors, respectively. In the first, the current which produces the magnetic field is independent of the current which flows through the plasma. In the second, the plasma current induces its own magnetic field and the interaction of these two gives rise to the accelerating force. In the third, the plasma is not subject to an external applied voltage, so it does not transmit current. The accelerating force is due to the interaction of an externally applied unsteady magnetic field with the circulating currents induced within the plasma by this field.

Yet other thrusters can be labelled electrodynamic and include magnetoplasmadynamic (MPD) and helicon thrusters. In MPD thrusters, a current along a conducting bar creates an azimuthal magnetic field that interacts with the current of an arc that runs from the point of the bar to a conducting wall. The resulting Lorentz force[2] has two components:

- Pumping: a radially inward force that constricts the flow.
- Blowing: a force along the axis that produces the directed thrust.

The principle of the helicon thruster is similar to the pulsed-plasma thruster: a traveling electromagnetic wave interacts with a current sheet to maintain a high $J \times B$ force on a plasma moving along an axis. This circumvents the pulsed-plasma thruster's problem of the force falling off as the current loop gets larger. The traveling wave can be created in a variety of ways, and a helical coil is often used. The plasma and coil of a helicon device appear below.

(Text adapted from the University of Wisconsin-Madison, Fusion Technology Institute's website and Mechanics and Thermodynamics of Propulsion, Hill and Peterson)

---

1 A useful working definition of a *plasma*, taken from F.F. Chen, **Introduction to Plasma Physics**, is "A plasma is a quasineutral gas of charged and neutral particles which exhibits collective behavior."
Quasineutral means that the number of positive and negative charges are very nearly equal.
Collective behavior means that the state of the plasma in regions somewhat distant from the point of interest may affect the behavior.

2 In physics, the Lorentz force is the force exerted on a charged particle in an electromagnetic field. The particle will experience an electric force qE and a magnetic force $qv \times B$.

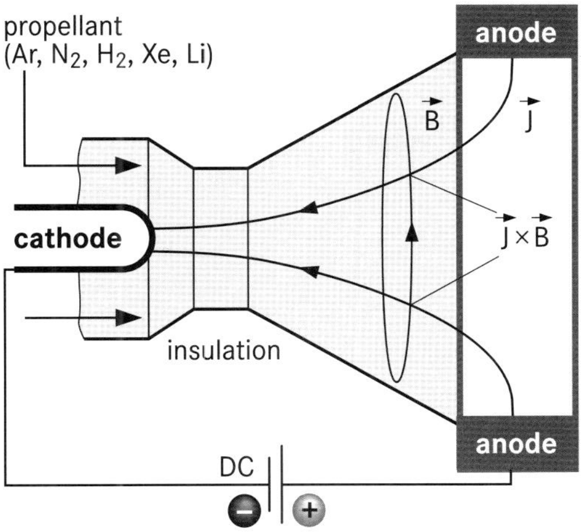

Fig. 9.3 Cross-section of a helicon thruster

9.1.1 Vocabulary

Translate the following vocabulary into German

English	German
fundamentally	
performance	
propellant, fuel	
thence (rare) = therefore	
essentially	
inertial	
charged	
conductive	
thruster	
fission reactor	
nozzle	

English	German
incidental	
azimuthal	
constrict	
circumvent a problem	

9.1.2 Questions

1) Name two benefits and two drawbacks of electrical and electromagnetic rockets compared to chemically propelled rockets.

2) Why are electrical rockets suitable for space missions?

3) Draw a diagram showing the relationship between the various types of electrical rockets.

9.2 Hall Effect Thrusters

Hall effect thrusters have been classed as both electrostatic and electromagnetic propulsion systems. An electrostatic field accelerates the ions in the propellant stream, but that field is, to a large extent, produced by the actions of plasma electrons interacting with a magnetic field, giving both classes a claim to the technique.

In Hall effect thrusters, also referred to as closed drift thrusters, a heavy gas, xenon in most implementations to date, is ionised and accelerated by an electric field. An electron current passing through and being impeded by a magnetic field establishes the acceleration field. The precessing electrons in this magnetic field follow a closed drift path, giving rise to one of the names for this technique. The flow of the drifting electrons, which is perpendicular to both the applied magnetic field and the current flow in the plasma, is due to the Hall effect, the source of the other name for this type of thruster.

The mechanical configuration of a Hall effect thruster is an annular discharge chamber in which a radial magnetic field is established between an inner, cylindrical ferromagnetic pole piece and an outer, ferromagnetic ring. The chamber is closed at one end, where an annular anode is situated. The other end of the chamber is open and forms the exit path for the accelerated ions. Outside the chamber, beyond the exit for the ion stream, is an external cathode. As the electrons emitted by the cathode move towards the anode under the influence of the applied electric field, the radial magnetic field results in a force acting on them in a direction perpendicular to the plane containing the two fields, which causes them to drift in an azimuthal direction. This azimuthal drift results in a Hall current and impedes the progress of the electrons towards the anode.

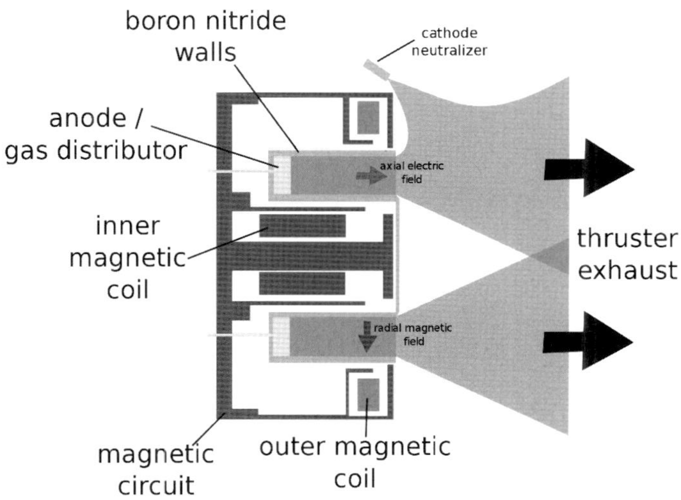

Fig. 9.4 Hall Thruster. Hall thrusters are largely axially symmetric. This is a cross-section containing that axis.

Hall/closed drift thrusters are divided into two types, stationary plasma thruster (SPT) and thruster with anode layer (TAL), based both on geometrical and material differences in the discharge channels of the two types, and on the different physical processes that take place within the discharge plasma.

In the stationary plasma thruster, the acceleration channel length is comparable to or greater than the diameter of the discharge chamber and the walls of the chamber are composed of a dielectric, ceramic material (usually boron nitride) that serves to protect the magnet poles from plasma erosion. The thruster with anode layer has a very short acceleration channel and the discharge plasma is positioned downstream of the thruster exit plane, obviating the need for protecting the magnet poles.

Although the two types of thruster are similar in layout and operating principle, the different materials and mechanical configurations used for the acceleration channels greatly change the underlying physics of operation of the thruster variants.

(Source: European Space Agency's website)

9.2.1 Vocabulary

Match the following words with the correct definitions.

1) to a large extent
2) implementation
3) impeded
4) precessing
5) annular
6) discharge
7) emitted
8) radial
9) erosion
10) obviating the need

a) .... the carrying out of a plan, idea or system

b) .... said when the tilt of an object's axis goes around in a circle in the opposite direction that the object is spinning

c) .... sent something out into the air, especially gas, light or heat

d) .... doing something differently in order to get rid of a problem or unwanted necessity

e) .... mainly

f) .... said of lines that emanate from the centre of a circle

g) .... the wearing away of a material

h) .... shaped like a ring

i) .... hindered, obstructed

j) .... the release of a substance

9.2.2 Comprehension Questions

1) Why have Hall effect thrusters been classified as both electrostatic and electromagnetic propulsion systems?

....

....

....

....

....

2) What are the differences between stationary plasma thrusters (SPTs) and thrusters with an anode layer (TALs)?

....

....

....

....

....

3) Label the parts in the figure below whose names are missing.

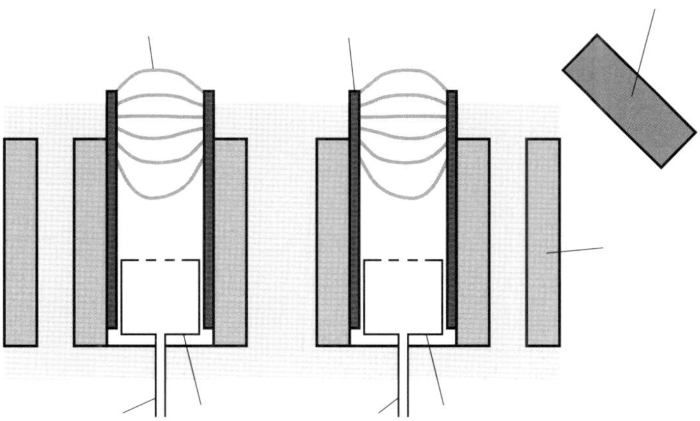

Fig. 9.5 Schematic cross section of stationary plasma thruster (used in SMART-1, ESA's space probe for moon exploration

9.3 Comparison of Propulsion Technologies

Examine the table below and summarize in approximately half a page the benefits of an electric propulsion system like the one used on SMART-1 compared to conventional chemical propulsion systems.

	Small monopro-pellant thruster (chemical)	Fregat Main Engine (SR.92M) (chemical)	SMART-1 Hall Effect Thruster (PPS-1350)
Propellant	Hydrazine	Nitrogen tetroxide / unsymmetrical dimethyl hydrazine	Xenon
Specific Impulse (s)	200	320	1,640
Thrust (N)	1	1.96×10^4	6.80×10^{-2}
Thrust time (s)	1.66×10^5	877	1.80×10^7
Thrust time (h)	46	0.24	5,000
Propellant con-sumed (kg)	52	5,350	80
Total Impulse (Ns)	1.1×10^5	1.72×10^7	1.2×10^6

(Source : European Space Agency)

9.4 Discourse Markers

Discourse markers are words or expressions used to link ideas together and show the relationship between them. Although the variety of discourse markers in everyday speech is somewhat limited, in academic writing there are numerous possibilities.

Match the following discourse markers with their most common equivalents:

And/More Information

But/In Contrast

For example

Because

As a result

In conclusion

to summarise	in order to	also	
in addition	on the other hand	thus	furthermore
so	resulting in	additionally	due to
moreover	not only... but also	though	whereas
therefore	such as	instead	because of
however	although	that is the reason why	on the contrary
consequently	in comparison	hence	since
unlike	as opposed to	for instance	as well as
in summary	despite	yet	

Important Points

The following two points are important to remember:

a) Certain discourse markers or linking words are only followed by a noun or noun group, e. g. *such as, because of, resulting in, as well as, unlike, due to*

Example

Unlike jet engines, rocket engines can only be used in space.

b) With certain discourse markers, it is necessary to create a new sentence when linking two ideas, for example with *on the other hand, however, consequently, moreover.*

Examples

Composite materials have revolutionized airplane design. Moreover, they have enabled airlines to lower their fuel costs.

Space agencies are eager to send humans back to the Moon. On the other hand, they are aware that operating space probes like SMART-1 is a more cost-efficient alternative.

(The following punctuation would be incorrect in English: *Space agencies are eager to send humans back to the Moon, on the other hand, they are aware that operating space probes like SMART-1 is a more cost-efficient alternative.)*

10 Orbits and SMART-1

Fig. 10.1 SMART-1 en route to the Moon using ion propulsion

10.1 SMART-1's Lunar Mission

SMART-1 was the first of a series of 'Small Missions for Advanced Research in Technology' designed to test key technologies for future spacecraft. It was also Europe's first mission to the Moon. The European Space Agency's SMART-1 was one of three payloads on Ariane Flight 162. The generic Ariane 5 lifted off from the Guiana Space Centre, Europe's spaceport at Kourou, French Guiana, at 2014 hrs local time (2314 hrs GMT) on 27 September (01:14 Central European Summer time on 28 September).

Listening Comprehension

Watch the introductory video about ESA's mission to the Moon ("Smart-1 en route to the moon") and answer the questions below. Enter the web address listed below into your Internet browser in order to gain access to the video.

http://www.esa.int/esa-mmg/mmg.pl?mission=SMART-1&type=V

1) What does SMART stand for?

2) When was SMART-1 launched? How was it launched into orbit?

3) One purpose of the SMART-1 mission is to carry out tests of its state-of-the-art technology. Name two such tests.

4) SMART-1's mission also includes carrying out scientific tests of the Moon. What are they?

5) Describe one theory about the origin of the Moon.

10.2 Orbits

10.2.1 Preliminary Questions

Working with a partner, see if you can answer the questions based on your previous knowledge of space engineering.

1) What are lunar resonances?

....

....

....

2) What does Lagrange Point No. 1 (L1) denote?

....

....

....

3) What typically happens to a spacecraft during a planetary fly-by?

....

....

....

4) How would you define in English what semi-major and semi-minor axes are?

....

....

....

5) What do think the difference is between a geosynchronous and a geostationary orbit?

....

....

....

10.2.2 Vocabulary

Exercise A

The following is a definition of geosynchronous and geostationary orbits. Use the words given to fill in the blanks.

travels	circles	adjusted
sidereal (x 2)	equator	pole
technically	sunset	inclination
direction	accounts	

A geosynchronous orbit is any orbit that .... the earth in exactly one day. In this situation "one day" refers to a .... day, which is slightly shorter than the sunrise/............ definition we usually use, and which .... for more factors than just the rotation of the earth. Furthemore, a geosychronous orbit can .... orbit in any .... around the earth.

A satellite in geostationary orbit also circles the earth in exactly one .... day and is, by definition, geosychronous. But in the case of a geostationary orbit the orbit's angle and direction are .... so that it is directly over the .... (............ is 0) and spinning around the earth in the same direction as the earth itself spins. Because the satellite now turns around the earth's .... just like the earth does and .... at the same speed and in the same direction as the earth, the satellite and the earth spin together and the satellite stays over the same place all the time.

(Source: Wikipedia)

10.2.3 Comprehension Questions

After reading the following text "A spiral pathway to the Moon", go back and review the answers you gave to the preliminary questions.

10.2.4 A spiral pathway to the Moon

The Ariane-5 launcher placed SMART-1 into an elliptical orbit around the Earth. Under the control of the European Space Operations Centre (ESOC) in Darmstadt, Germany, on two days per week, repeated burns of the ion engine changed the ellipse into a circle, gradually expanding it into a spiral.

Month by month, the Moon revolves around its own orbit, 350,000 to 400,000 kilometres from the Earth. As SMART-1 gained distance from the Earth, its speed slackened. When 200,000 kilometres out, the spacecraft began to feel significant gravitational tugs from the Moon as it passes by.

Mission controllers then, bringing in a new era of space navigation, for the very first time, used the sustained thrust of electrical propulsion jointly with manoeuvres under gravity. Isaac Newton knew nothing of such tricks, and ESA's experts had to invent fresh mathematics for figuring out the best orbits.

The tug of the Moon's gravity at first helped to widen the spiral orbit, in regular encounters called 'lunar resonances'. By the time SMART-1 passed within 60,000 kilometres of the Moon, the effect of gravity was much more pronounced, in encounters known as 'lunar swing-bys'.

At a crucial stage in the journey, called lunar capture, SMART-1 passed through an invisible doorway in space at Lagrange Point No. 1, or L1 for short. As first noted by the mathematician Joseph-Louis Lagrange in 1772, the gravitational effects of the Moon and the Earth are in balance at L1, 50,000 to 60,000 kilometres out from the Moon on the earthward side.

Beyond L1, SMART-1 flew over the lunar north pole, aiming at a point of closest approach above the south pole, so achieving a wide polar orbit around the Moon. During the weeks following its capture by the Moon, SMART-1's ion engine gradually reduced the size and duration of this orbit, to improve the view of the lunar surface.

(Source: European Space Agency)

10.2.5 Translations

Translate the following sentence fragments from the preceding text into German.

a) its speed slackened ...

b) the sustained thrust of electrical propulsion ...

c) gravitational tugs from the Moon ...

d) the effect of gravity was much more pronounced ...

Fig. 10.2 Test firing of an ion engine

10.3 Ion Electric Propulsion on SMART-1

The planetary objective selected for the SMART-1 mission was to orbit the Moon for a nominal period of six months. It was the first time that Europe sent a spacecraft to the Moon. The project aimed to have the spacecraft ready by late August 2003 for launch as an Ariane-5 auxiliary payload.

Listening Comprehension

Watch a further video about ESA's mission to the Moon ("Smart-1: driving an ion engine in space") and answer the questions below. Enter the web address listed below into your Internet browser in order to gain access to the video.

http://www.esa.int/esa-mmg/mmg.pl?mission=SMART-1&type=V

1) What is the propellant used on SMART-1? What provides the electricity?

....

....

....

....

2) Why did ESA choose to use an electric propulsion system on SMART-1?

....

....

....

....

3) It is claimed in the video that using an electric propulsion system for longer space missions, for instance to Mercury, could save two to four years. Explain how this is possible, given that conventional propulsion systems (i.e. chemical propulsion) deliver much higher accelerations and therefore are quicker for trips to the Moon.

....

....

....

....

Explain to a fellow student by means of a diagram how the SMART-1 spacecraft will reach the moon. Be sure to use the expressions **intermittently** and **elliptical geostationary transfer orbit** in your explanation.

10.4 Clarifying Information

When dealing with highly technical topics, we are often confronted with situations where we do not immediately understand the information being conveyed. In such situations, we seek to clarify the situation in order to check our understanding.

The following are some helpful expressions used to clarify information in English:

- What do you mean by ...?
- I'm not sure I understand that bit about ...
- I didn't quite get what you said about ...
- Would you mind repeating that part about ...?
- I'm not sure I follow you.
- In other words, you mean ...
- If I could summarize what you just said ...
- Do you mean to say that...?
- I'm afraid I don't quite follow you. Could you repeat that?
- So, in other words, ...
- If I heard you correctly, you said ...
- So, what you're trying to say is that ...

10.4.1 Speaking Exercise

Working with a partner, divide the following two texts up between you. Read your text and summarize it orally for your partner. Your partner should listen carefully and ask for clarification when needed using the expressions given above.

Text A: What do all the instruments do?
Multinational teams of scientists and engineers are conducting ten different investigations coordinated by a Science and Technology Operations Centre. The instrument teams are led by Principal Investigators from Finland, Germany, Italy, Switzerland and the United Kingdom. All ESA member countries are taking part, providing Co-Investigators for various experiments.

Testing new techniques
EPDP and **SPEDE**: Designers of future solar-electric spacecraft want to know how SMART-1's ion engine performs, what side-effects it has, and whether the spacecraft interacts with natural electric and magnetic phenomena in the space around it. Possible problems include deflection of the ion engine's thrust direction, erosion of surfaces, short-circuits by sparks, interference with radio signals, and accumulating dust. The main on-board instruments monitoring these effects are EPDP and SPEDE.

KaTE and **RSIS**: Small changes in SMART-1's motion reveal the precise thrust delivered by the ion engine. Like police radars used to catch speeding motorists, RSIS employs the Doppler effect, to see how the speed alters the wavelength of radio pulses. It uses the very short radio waves of KaTE. The primary purpose of KaTE is to demonstrate the next generation of radio links between the Earth and far-flung spacecraft. Microwaves of the Ka band, around 9 millimetres in wavelength, can be focused into relatively narrow beams by the small dish antennas available on the spacecraft.

Laser Link is another communications experiment. ESA already has laser links with telecommunications satellites from an optical ground station on Tenerife in Spain's Canary Islands. Aiming the beam becomes much more difficult if, like SMART-1, the spacecraft is far away and moving rapidly. The hope is that the on-board camera AMIE will see Tenerife aglow with laser light.

OBAN: Future spacecraft will be more self-reliant in guiding themselves along pre-defined paths towards distant destinations. OBAN is evaluating a computer technique for on-board autonomous navigation. It uses the bearings of stars seen by SMART-1's star trackers, and the Earth, Moon, and possibly asteroids seen by the AMIE camera.

Observing the Moon and the Sun
AMIE, SIR and **D-CIXS** Different kinds of visible and invisible light coming from the lunar surface provide clues about its chemical composition and geological history. The ultra-compact electronic camera, AMIE, is surveying the terrain in visible and near-infrared light. An infrared spectrometer, SIR, charts the Moon's minerals. An X-ray telescope, D-CIXS, identifies key chemical elements in the lunar surface. The major scientific goals are described in the following pages.

XSM: The D-CIXS measurements could be confused by variations in solar X-ray emissions, which depend on how stormy the Sun is at the time. So SMART-1 monitors the solar X-rays with its XSM instrument. XSM also makes its own independent study of solar variability.

SPEDE: Like a ship at sea, the Moon leaves a wake in the solar wind - the non-stop stream of charged particles and associated magnetic fields coming from the Sun. The SPEDE electrical experiment observes this effect at close quarters.

RSIS: With help from the KaTE microwave system and the AMIE camera, the RSIS radio experiment demonstrates a new way of gauging the rotations of planets and their moons. It should be able to detect a well-known nodding of the Moon, which slightly tilts first its north pole and then its south pole towards the Earth.

Instruments and techniques being tested in examining the Moon from SMART-1 will later help ESA's BepiColombo spacecraft to investigate the planet Mercury.

(Source: www.esa.int.)

Text B: Masterpieces of miniaturisation

The SMART-1 spacecraft spans 14 metres with its solar panels extended, but otherwise everything for propulsion, communications, housekeeping and instrumentation fits into a cube just 1 metre across.

Propulsion by an ion engine is not the only innovative technology on SMART-1. Its solar panels use an advanced type of gallium-arsenide solar cells in preference to the traditional silicon cells. And it is testing new communications and navigational techniques.

Out of a total mass at launch of 370 kilograms, the payload available for a dozen technological and scientific investigations is 19 kilograms. Like other components of the spacecraft, the scientific instruments use state-of-the-art concepts and methods of miniaturization to save space and economise on mass. For example the X-ray telescope D-CIXS makes a cube just 15 centimetres wide and weighs less than 5 kilograms.

10.4.2 SMART-1 mission overview

Purpose:	Flight testing electric propulsion and other deep-space technologies, while performing scientific observations of the Moon.
Spacecraft:	One cubic metre, 370 kilograms. Solar panels span 14 metres when deployed and provide 1.9 kWatts of power.
Scientific payload:	19 kilograms
Launch:	August 2003 from Kourou, French Guiana on a shared Ariane-5 ride to geostationary-transfer orbit (GTO).
Orbit:	16-month transfer orbit from GTO to lunar orbit insertion, then polar elliptical operational orbit, ranging from 300 to 10,000 kilometres in altitude above the Moon.

Ground stations:	ESA network stations around the world, operating 8 hours twice a week.
Mission lifetime:	2 - 2.5 years
Prime contractor:	Swedish Space Corporation, Solna, Sweden

(Source: www.esa.int.)

10.5 Cause and Effect

10.5.1 Exercise A

Place the following verbs under the appropriate column according to whether they express cause or effect.

to cause
to result from
to affect
to be the cause of
to be produced by
to be affected by
to result in
to lead to
to be caused by
to produce
to be the consequence of
to be due to
to be the result of
to bring about

Cause	Effect

The following prepositional phrases can also be used to show a causal relationship between two events or phenomena

- because of
- owing to
- on account of
- as a result of
- due to[1]

10.5.2 Exercise B

Link the following pairs of sentence fragments in a complete sentence in order to explain the cause and effect relationship between them both. You might need to alter some of the words given slightly and add a few of your own.

a) Increased velocity of airflow along the top of an airfoil lift

b) Higher fuel consumption greater drag.

c) The angle of attack lift

d) Fuel combustion thrust.

e) The pitching motion deflection of an aircraft's elevator.

f) The weakening of composite materials ageing and stres

g) A space elevator lower launch costs.

1 Some linguists consider „due to" to be an adjectival phrase and therefore not interchangeable with „because of" and prepositional synonyms. However, in modern everyday English, this distinction does not hold.

11 Commercial Space Flight

Fig. 11.1 SpaceShipOne

11.1 Preliminary Discussion Questions

In groups of two or three, discuss the following topics and be prepared to summarize the points discussed for the entire class.

1) What is the future of commercial space flight? Will it continue to be a vain pursuit of the very wealthy or do you think it could be a feasible commercial venture in the medium to long term?

2) Do you think space agencies like NASA and ESA will turn to commercial space flight, through joint ventures for example, as a means of raising additional revenue and/or improving and testing new technologies?

3) If you were to design a vehicle for private space flight, describe how the vehicle would be launched and what kind of propellant would be used? Are there any other design specifications that you would consider?

4) Compare and contrast liquid, solid and hybrid propellants? Which would you choose for your vehicle? Why?

Liquid	Solid	Hybrid

11.2 Commercial Space Flight: Listening Comprehension

Listen to the radio broadcast from National Public Radio dealing with commercial space flight and answer the following questions. Enter the following web address into your browser to gain access to the broadcast.

http://www.npr.org/templates/story/story.php?storyId=3821627

1) What are the conditions of the X prize?

....

....

....

....

2) Describe the appearance of the Canadian space vehicle–dimensions, colour, features.

....

....

....

....

3) Describe the propulsion system and how the vehicle works.

....

....

....

....

4) What is the vehicle's name?

5) How much has been spent by the Canadian team so far?

6) How does the Canadian vehicle differ from the American one?

7) Describe the Canadian group's next project.

11.2.1 Exercise A

Compare the Canadian group's commercial spacecraft with the three designs described in the following texts. How are they similar? How are they different?

Create a table to provide a quick overview of the similarities and differences. You will need to come up with your own criteria.

Criterion	Canadian Space Vehicle	Rocketplane XP	SpaceShipTwo	Xerus

11.2.2 Exercise B

Which design is the most feasible? Justify your decision in half a page.

Rocketplane XP

The **Rocketplane® XP** is a suborbital space plane under development by Rocketplane Limited, Inc. The XP is based on a heavily-modified Learjet 25 airframe, with changes including the addition of a delta wing and V-tail. The vehicle is powered by two jet engines and a rocket engine enabling it to reach suborbital space. The XP is designed to operate from existing spaceports in a manner consistent with established commercial aviation practices.

The Rocketplane® XP, carrying a pilot and three commercial participants (passengers), will take off from a runway using a jet engine just like regular aircraft and ascend to about

30,000 feet. At this point, a reusable rocket engine powers the XP to altitudes of over 100 km (62 miles). It will give participants a sensation of weightlessness for three to four minutes. These participants will experience the same view previously only known to a select few astronauts. The XP will then re-enter Earth's atmosphere and land at the same spaceport under conventional jet power.

On January 24, 2006 Rocketplane Limited announced a Space Act agreement with NASA Johnson Space Center for the loan of a Rocketdyne RS-88 rocket engine for three years, for use in flight tests of the XP vehicle.

In addition to the XP, Rocketplane Limited is developing multiple generations of spaceplanes to serve the civilian and military markets based on successful United States Department of Defense (DOD) and NASA awards as well as continued growth in the space tourism market.

The XP will operate from the Oklahoma Spaceport near Burns Flat, Oklahoma.

(Source: Wikipedia)

Xerus

This craft is the next vehicle step beyond the EZ-Rocket. The details of configuration will change before the design is finished. At the top of our list of priorities for the suborbital vehicle is low operational cost. It will be powered by XCOR rocket engines, scaled up from the EZ-Rocket engines, which have already been shown to have long lifetime and operation costs.

A cluster of main engines will propel the vehicle from the runway to about 65 km altitude, after which it coasts to 100 km during a typical sounding rocket flight. Initial flight testing will retain a propellant reserve so that the pilot can restart the engines to reach the airport, or perform a go-around if he chooses. XCOR plans a flight test program of at least 20 flights, each incrementally expanding the operational envelope.

Maximum speed is about Mach 4, which is achieved high in the atmosphere where aerodynamic heating is much less of a problem than it is for airplanes that cruise supersonically lower in the atmosphere. Thermal protection is correspondingly easier.

The wings provide for runway takeoffs and landings as well as maneuvering in the atmosphere. Small rockets, based on our 50 lbf thruster, will be used for attitude control outside the atmosphere.

(Source: www.xcor.com)

SpaceShipTwo

Scaled Composites' **Model 339 SpaceShipTwo** (SS2) is a suborbital space plane for carrying space tourists, under development by The Spaceship Company, a joint venture between Scaled Composites and Sir Richard Branson's Virgin Group, as part of the Tier 1b program. The spaceship was officially unveiled to the public on Monday, 7 December 2009, at the Mojave Air and Spaceport in California. The Virgin Galactic spaceline plans to operate a fleet of five of these craft in passenger-carrying private spaceflight service starting no earlier than 2011. A ticket costs $200,000 USD.

Design and development

The SpaceShipTwo project is based in part on technology developed for SpaceShipOne as part of the Scaled Composites Tier One program, funded by Paul Allen. The Spaceship Company licenses this technology from Mojave Aerospace Ventures, a joint venture of Paul Allen and Burt Rutan, the designer of the predecessor technology.

SpaceShipTwo will carry passengers to space; the capacity will be eight people: six passengers and two pilots. The apogee of the new craft will be approximately 110 km (68 miles) and in the thermosphere, 10 km (6 miles) higher than both the SpaceShipOne target (though the last flight of the SpaceShipOne reached 112 km), and the Kármán line. SpaceShipTwo will reach 4,200 km/h (2,600 mph), using a single hybrid rocket motor, which goes by the name RocketMotorTwo. It will launch at 15,200 m (50,000 ft) from its mother ship, White Knight Two, and go supersonic within 8 seconds. After 70 seconds, the rocket motor cuts out and the spacecraft will coast to its peak altitude. SpaceShipTwo's crew cabin is 3.66 m (12 ft) long and 2.28 m (7.5 ft) in diameter. The wing span is 8.23 m (27 ft), the length is 18.29 m (60 ft) and the tail height is 4.57 m (15 ft).

SpaceShipTwo uses a feathered reentry system, feasible due to the low speed of re-entry – by contrast, space shuttles and other orbital spacecraft re-enter at orbital speeds, closer to 25,000 km/h (16,000 mph), using heat shields. It is designed to reenter the atmosphere at any angle.

SpaceShipTwo will decelerate through the atmosphere, switching to a gliding position at 80,000 ft and will take 25 minutes to glide back to the spaceport. Once the passengers return, their passports will receive a spaceflight stamp and they will be awarded their astronaut wings at an awards ceremony. They will also be able to relive their flight from footage filmed during the spaceflight.

SpaceShipTwo and its new carrier aircraft, White Knight Two, are roughly twice the size of the first generation spacecraft SpaceShipOne and mothership White Knight that won the Ansari X Prize. SpaceShipTwo will have 43 cm and 33 cm (17 and 13 in) diameter windows for the passengers' viewing pleasure, and all seats will recline back during landing to decrease the discomfort of G-forces. Reportedly, the craft can land safely even if "catastrophic damage" occurs during the flight.

(Source: Wikipedia)

Fig. 11.2 SpaceShipTwo

11.3 Translation: SpaceShipOne

After reading about SpaceShipTwo, try and translate into English the following passage taken from Wikipedia relating to some of the technical details of the vehicle's predecessor.

Grundsätzlich ist *SpaceShipOne* nicht für Flüge in eine Erdumlaufbahn ausgelegt, da seine Maximalgeschwindigkeit nur etwa ein Siebtel der dafür nötigen ersten kosmischen Geschwindigkeit beträgt.

Das Flugzeug erreicht während der 65 Sekunden dauernden Brennzeit des Triebwerkes eine Geschwindigkeit von Mach 3,5. Es hat ein von der US-amerikanischen Firma SpaceDev gebautes Hybrid-Raketentriebwerk, dessen Treibstoff aus einem festen Energieträger (HTPB) - einer Art Reifengummi - und einem flüssigen Sauerstoffträger besteht. Es ist teilweise wiederverwendbar. Der ballistische Raumflug dauert etwa dreieinhalb Minuten, dann wird wieder die obere Schicht der Atmosphäre erreicht. Beim Start wird das Flugzeug von einem Träger, dem sogenannten White Knight, in eine Flughöhe von etwa 15 km geschleppt und dort ausgeklinkt. Daraufhin zündet das Raketentriebwerk.

(Adapted from Wikipedia)

....

....

....

....

....

....

....

11.4 Respectively/*Beziehungsweise*

Respectively

a) **Meaning/Use**:

The word "respectively" is used at the end of a sentence to show the relationship between two pairs of items, i.e. which item goes with which item.

Examples:

"John and Mary like ice cream and hot dogs respectively."

The use of "respectively" here indicates clearly that John likes ice cream and Mary hot dogs. If we didn't use it in this sentence, it would be ambiguous and suggest that they both like both items.

"The head offices of NASA and ESA are located near the centre of power and political decision-making, in this case in Washington, D.C. and Paris, France respectively."

b) **Position**: 'Respectively' is placed at the end of the clause. This is different from German!

c) **Translation:** *beziehungsweise*

„Der Hauptsitz von NASA und ESA befindet sich in der Nähe des Machtzentrums und der politischen Entscheidungsträger, sprich in Washington, D.C. bzw. in Paris, Frankreich."

d) Please Note:

Beziehungsweise can be translated in three ways.

i) **respectively** (see above)

ii) **or rather**

e.g. *„Die Airbus-Fabrik liegt in Hamburg bzw. im Stadtteil Finkenwerder."*

The Airbus plant is located in Hamburg or rather in Finkenwerder.

iii) **or** (this is the most common)

e.g. *„Firma X bzw. die Firma Y wird die neuen Flugzeuge warten."*

Company X or company Y will service the new planes.

12 International Space Station

Fig. 12.1 Picture of the ISS, taken on May 23, 2010

12.1 The International Space Station

12.1.1 Overview of International Space Station

Description and Purpose

The ISS supports a variety of pure and applied research for the US and its international partners and will play a critical role in supporting the Vision for Space Exploration. The ISS provides a unique, continuously operating capability: to develop medical countermeasures for long-term human space travel, to develop and test technologies and engineering solutions in support of exploration, and to provide ongoing practical experience living and working in space. More than four times as large as the Russian Mir when completed, the ISS will be the largest, most complex international scientific project in history and Boeing's largest adventure into space to date.

Boeing is NASA's prime contractor to design, develop, integrate, test and deliver the US built elements of the ISS. The US elements include three connecting modules, or nodes; a laboratory module; truss segments; four solar arrays; three mating adapters; a cupola; and an unpressurized logistics carrier. The various systems developed by Boeing include thermal control; life support; guidance, navigation and control; data handling; power systems; and communications and tracking.

Boeing also integrates vehicle elements provided by the international partners. International partner components include: a Canadian-built 55-foot-long robotic arm used for assembly and maintenance tasks on the space station; a pressurized European laboratory called Columbus and logistics transport vehicles; a Japanese laboratory called Kibo with an attached exposed exterior platform for experiments as well as logistics transport vehicles; and two Russian research modules; an early living quarters called the Service Module with its own life support and habitation systems, logistics transport vehicles; and Soyuz spacecraft for crew return and transfer.

Boeing is completing the US on-orbit segment of the ISS and providing vehicle-sustaining engineering. The company also prepares every ISS US component for space flight at the Space Station Processing Facility at Kennedy Space Center, FL. About 1,850 Boeing people across the country support the space station. Boeing employees in Houston, Huntsville, AL, Huntington Beach, CA, and Kennedy Space Center are working with NASA on the ISS project.

Table Characteristics of the ISS upon Completion

Characteristic	Data
Solar Array Wingspan (port to starboard)	356 feet (108.5 m)
Length (pressurized section)	167 feet (51.0 m)
Solar Array Wingspan (tip to tip)	239 feet (72.8 m)
Integrated Truss Length	310 feet (94.5 m)
Mass	about 1 million lb (453,592 kg)
Operating Altitude	220 nmi average (407 km)
Inclination	51.6 degrees to the equator
Atmosphere Inside	14.7 psi (101.36 kilopascals)
Pressurized Volume	46,000 cubic feet
Computers to control ISS	52
Robotic Arm	55-foot robot arm assembly that can lift 220,000 pounds
Power Generation	110 kilowatts
Crew Size	Six

(Adapted from Boeing/ISS_overview)

12.1.2 Comprehension Questions

With a partner, try and answer any questions you may have about the text, including the difficult vocabulary listed below. The paragraph where the vocabulary can be found in the text is indicated in brackets after each word or expression.

- truss segment (2)
- cupola (2)
- mating adapters (2)
- logistics carrier (2)
- attached exposed exterior platform (3)
- vehicle-sustaining engineering (4)
- on-orbit segment (4): What other prepositions can you use with orbit?

  NASA launched the satellite .... orbit.

  The satellite is .... orbit around the Earth.
- port (in previous table)
- starboard (in previous table)
- solar array (in previous table)
- inclination (in previous table)

What is the purpose of the many semi-colons (;) in paragraphs two and three?

Fig. 12.2 Close-up view of the ISS after assembly work has been completed

12.2 Ad Lib Elaboration: Zarya and Unity Modules

Zarya Control Module

Take a look at the notes below dealing with the Zarya control module for the ISS. Summarize them to your partner orally, elaborating on the information to ensure that your explanation flows smoothly. Your partner should do the same for the Unity Control Module on p. 137.

Name

- means "sunrise" in Russian
- also known as the FGB (Functional Cargo Block)

Origin/History

- launch: Nov. 20, 1998, Kazakhstan, Russian Proton rocket
- built by Khrunichev State Research and Production Space Centre in Moscow
- first component of ISS

Purpose/Role

- station's initial propulsion and power
- orientation control and communications power
- after Service Module is launched → storage capacity and external fuel tanks

Quick Facts

- Length (end-to-end): 41.2 feet
- Width (at widest point): 13.5 feet
- Mass in orbit: 44,088 pounds
- Lifetime: 15 years
- Inclination of orbit: 51.6 degrees
- Orbit at rendezvous: 240 statute miles (1 mile = 1.6 km) circular
- Docking Ports: Soyuz (piloted), Progress re-supply spacecraft (unpiloted)
- Power = 6 kW

  solar arrays + 6 nickel-cadmium batteries
- 16 fuel tanks (6 tons of propellant)
- 24 large steering jets, 12 small ones (attitude control system)

(Adapted from www.shuttlepresskit.com)

Fig. 12.3 Zarya (1998)

Fig. 12.4 Zarya and Unity (1998)

Unity Control Module

Take a look at the notes below dealing with the Unity control module for the ISS. Summarize them to your partner, elaborating on the information to ensure that your explanation flows smoothly.

History/Origin

- first US-built component
- also known as Node 1
- launch method: Space Shuttle Endeavour
- manufacturing facility: Boeing in Alabama
- mated with already orbiting Zarya control module

Shape/Features

- six-sided, six berthing ports for future modules
- 15 feet (diameter), 18 feet (length)
- made of aluminium
- 50,000 mechanical items
- 121 internal and external electrical cables
- six miles of wire
- data, voice and low data rate video communications system for Mission Control in Houston

Purpose/Role

- life support systems, environmental control, electrical and data systems → work and living areas

Pressurized Mating Adapters

- two, conical docking adapters
- attached to each end
- berthing/docking mechanisms for Space Shuttle and Zarya
- one adapter → multiplexer-demultiplexers (MDMs) (computers for vehicle's command and control)

(Adapted from www.shuttlepresskit.com)

12.3 Systems Operations

12.3.1 Systems Operations of an ISS Module

The ISS is dedicated to performing research under microgravity conditions. Nevertheless, a substantial part of the available time has to be invested in securing and maintaining the necessary living and working conditions for the astronauts. Two of the three astronauts on board the ISS in 2007 had the sole task of performing maintenance on the ISS modules and repairing faulty parts. Only half of the working time of the third crew member was available for conducting experiments and for payload operations.

In addition to the work of the ISS crew, the flight control teams (FCTs) in the control centres provide an enormous amount of support to keep the ISS operational. In the control centres in Houston and Moscow, and since 2008 also in Munich and Tokyo, the FCTs work around the clock to operate the module assigned to their responsibility. The teams ensure that the module is working correctly, that the necessary maintenance is performed in time, that the environmental conditions for the astronauts are met, and that the payload operations can be performed.

12.3.2 System Operations of the Columbus Module

The main task of the FCT at the Columbus Control Centre (Col-CC) is to activate, command and monitor the Columbus module after its berthing with the space station.

Fig. 12.5 Extra-vehicular activity during the Columbus mission

The FCT, which is available 24 hours a day in three shifts during steady-state operations, has to ensure that the environmental control and life support system (ECLSS) is working correctly, and that the power supply and distribution are working properly and providing sufficient power to the subsystems and experiments. Additionally, the thermal subsystem for cooling the subsystems and experiments has to be monitored, and it must be assured that all data, video and audio connections are properly routed. In case of anomalies or faulty equipment the team has to perform countermeasures and provide a description of the failure to the engineering support team for analysis.

The engineering support team consists of experts from the manufacturers of the various subsystems of the Columbus module. Most of the team members reside at the Engineering Support Centres (ESCs) in Bremen and Turin, while a small team supports the FCT at Col-CC directly. They are informed by the FCT about failures in the module and provided with the necessary data. The ESC analyzes the anomaly, tries to reproduce the failure in simulators on the ground, and issues a recommendation on how to solve the problem. It is the task of the FCT to execute the recommended procedures in order to regain the full functionality of the module.

(Source: Handbook of Space Technology, Hanser/Wiley)

12.3.3 Collocations

A collocation is a series of words that appear together with a high statistical probability thereby forming one unit or expression. For example, the verbs "do, conduct and carry out" all collocate with the noun "research". However, "make research" is not a collocation in English, although it is grammatically correct. A further example taken from German will help illustrate what a collocation is.

Ich bin der .... Überzeugung...

In this case, "festen" or "vollen" would fit into the blank above forming a strong link with "Überzeugung". The adjectives "konkreten" and "soliden" would not form a collocation in German, as they would sound odd in this context.

Fig. 12.6 German astronaut Hans Schlegel performing his extra-vehicular activity during the Columbus mission

Exercise

What verbs collocate with the following nouns? The answers are found in the previous text.

conditions

maintenance

12.4 Auslautverhärtung / Terminal Devoicing

**The following is a simplified explanation of this phonological phenomenon and the references to German may vary from southern German dialects.*

Terminal devoicing is a phonological phenomenon that occurs in certain languages such as German and Dutch. When a voiced obstruent *(stimmhafter Geräuschkonsonant)*, for example v, b, d, g or z, appears at the end of a word in German it becomes voiceless.

e.g. *Rad* and *Rat* are pronounced the same

In English, however, there is a contrast between the phonemic pairs *bad* and *bat*.

Examine the following pairs in the table below. Try pronouncing the word pairs being careful to differentiate between the two items in each pair.

Voiced Consonants	**Voiceless Consonants**
/ v / e.g. believe, leave	/ f / e.g. belief, leaf
/ b / e.g. tab	/ p / e.g. tap
/ d / e.g. had	/ t / e.g. hat
/ g / e.g. bag	/ k / e.g. back
/ z / e.g. these, trans	/ s / e.g. this (the vowel is also pronounced differently here) trance

Credits

Every effort has been made to trace the owners of the copyright material contained within this textbook. In those rare cases where it has not been possible to contact the owners or where material has been overlooked, I would like to extend my sincere apologies.

The author would specifically like to thank the following copyright owners for their kind permission to use their materials:

www.hubschraubermuseum.de
New Scientist: www.newscientist.com
NASA: www.nasa.gov
DLR: Deutsches Zentrum für Luft- und Raumfahrt
Engineers Media, Paul Grad: www.engineersmedia.au
Bibliographisches Institut AG, excerpts from "How things work", Mannheim: 1967
European Space Agency: http://www.esa.int/esaCP/index.html
US Department of Transportation, Federal Aviation Administration: excerpts from Pilot's Handbook of Aeronautical Knowledge
Pearson Education, Inc.: excerpts from "Mechanics and Thermodynamics of Propulsion", Philip Hill and Carl Peterson
Planet Aerospace, GeraNova Bruckmann Verlagshaus GmbH: www.verlagshaus.de
University of Wisconsin, http://fti.neep.wisc.edu/~jfs/homepage.html, Prof. John Santarius, Department of Engineering Physics, Fusion Technology Institute

Chapter 1:

Fig. 1.3: DLR
Fig. 1.4, Fig. 1.5, Fig. 1.6 and Fig. 1.8: How things work: The universal encyclopedia of machines, Bibliographisches Institut Mannheim 1963
Fig. 1.9: BDLI (MTU Aero Engines)
Fig. 1.10: Airbus S.A.S.
Fig. 1.12: Airbus S.A.S.

Chapter 2:

Fig. 2.8: DLR
Fig. 2.9: NASA

Chapter 3:

Fig. 3.4: Pilot's Handbook of Aeronautical Knowledge; FAA, 4-8, 4-9
Fig. 3.5: JOURNEY IN AERONAUTICAL RESEARCH, Monographs in Aerospace History, Number 12

Fig. 3.6 and Fig. 3.8: Pilot‘s Handbook of Aeronautical Knowledge; FAA, 4-8, 4-9
Fig. 3.7: Wikipedia

Chapter 4:

Fig. 4.1: NASA Glenn Research Center
Fig. 4.2: NASA
Fig. 4.3: DLR
Fig. 4.4: ILA Berlin Air Show 2010 (BDLI)
Fig. 4.5: MTU Aero Engines GmbH München
Fig. 4.6 and Fig. 4.7: ILA Berlin Air Show 2010 (BDLI)
Fig. 4.8: Pratt & Whitney
Fig. 4.9: NASA Glenn Research Center
Fig. 4.10: DLR
Fig. 4.11: ESA
Fig. 4.12: NASA Photo
Fig. 4.13: DLR/Rolls Royce, supplemented

Chapter 5:

Fig. 5.1: NASA Scientific and Technical Information

Chapter 6:

Fig. 6.1: How things work, Bibliographisches Institut, pp. 560-561
Fig. 6.2 and Fig. 6.3: ILA Berlin Air Show 2010 (BDLI) (supplemented)
Fig. 6.4 , Fig. 6.5 and Fig. 6.6: How things work, Bibliographisches Institut, pp. 560-561
Fig. 6.7, Fig. 6.8, Fig. 6.9 and Fig. 6.10: BDLI
Fig. 6.11: DLR

Chapter 7:

Fig. 7.1 and Fig. 7.2: NASA
Fig. 7.3: ESA
Fig. 7.4 - Fig. 7.10: NASA

Chapter 8:

Fig. 8.1: NASA
Fig. 8.2: BDLI/EADS Astrium/ESA

Chapter 9:

Fig. 9.1: EADS/BDLI
Fig. 9.3: University of Stuttgart‘s Electric Space Propulsion website
Fig. 9.4: en.wikipedia.org
Fig. 9.5: ESA (website)

Chapter 10:

Fig. 10.1: ESA/DLR
Fig. 10.2: NASA

Chapter 11:

Fig. 11.1: NASA
Fig. 11.2: Wikipedia/Virgin Galactic/Mark Greenberg

Chapter 12:

Fig. 12.1–Fig. 12.6: NASA

Answer Key:

Fig. AK 7: wikimedia.org/wikipedia
Fig. AK 9: ESA Science & Technology

Selected Reference List

Websites

www.hubschraubermuseum.de
www.boeing.com
www.nasa.gov
www.dlr.de
www.wikipedia.org
http://www.esa.int/esaCP/index.html
www.leo.org

Technical/Scientific Magazines

New Scientist
Engine
Engineers Media
Aviation Week and Space Technology
Planet Aerospace
Journal of the British Interplanetary Society, specifically Vol. 59, No. 9 dealing with space elevators

Technical Books

1. *Aigner, Helmut:* Schlüssel zur englischen Fachsprache der Mathematik und Datenverarbeitung; R. Oldenbourg Verlag: Wien, 1975
2. *Brown, Gregory N.; Holt, Mark J.:* The Turbine Pilot's Flight Manual, 2nd Edition; Aviation Supplies and Academics, Inc.: Newcastle, Washington, 2001
3. *Evans, Julien:* How Airliners Fly: A Passenger's Guide; The Crowood Press Ltd., Marlborough, 2002
4. *Hall, Rex:* The International Space Station, From Imagination to Reality; British Interplanetary Society: London, 2002
5. *Johnson, Wayne:* Helicopter Theory; Dover Publications: New York, 1994
6. *Ley, Wilfried; Wittmann, Klaus; Hallmann, Willi:* Handbook of Space Technology; John Wiley & Sons Ltd: Chichester, 2009
7. Pilot's Handbook of Aeronautical Knowledge; US. Department of Transportation, Federal Aviation Administration, 2003

8. *Stinton, Darrol:* The Anatomy of the Aeroplane, Second Edition; Blackwell Publishing: Oxford, 1998

9. *Sutton, George P.; Biblarz, Oscar:* Rocket Propulsion Elements, Seventh Edition; John Wiley and Sons: New York, 2001

Dictionaries and Reference Materials

10. *Cescotti, Roderich:* Aerospace Wörterbuch: Aerospace Dictionary; Motorbuch Verlag: Stuttgart, 2002

11. *Downing, Douglas:* Dictionary of Mathematical Terms; Barron's: New York, 1987

12. How Things Work: The Universal Encyclopedia of Machines; Bibliographisches Institut: Mannheim, 1967

13. *Kumar, Bharat:* An Illustrated Dictionary of Aviation; McGraw-Hill: New York, 2005

14. Macmillan English Dictionary for Advanced Learners; Macmillan Education: Oxford, 2002

15. Macmillan Collocations Dictionary for Learners of English; Macmillan Education: Oxford, 2010

16. *McCarthy, Michael; O'Dell, Felicity:* Academic Vocabulary in Use; Cambridge University Press: Cambridge, 2008

17. *McCarthy, Michael; O'Dell, Felicity:* English Collocations in Use; Cambridge University Press: Cambridge, 2005

18. *Williamson, Mark:* The Cambridge Dictionary of Space Technology; Cambridge University Press: Cambridge, 2001

Answer Key

Chapter One: Aerodynamic Forces

1.1

boat: gravity or weight pulling the boat down; buoyancy pushing the boat up

rock: gravity or weight of the rock directed downwards; friction acting against the rock on the incline; force relative to the slope pulling the rock down the incline

aircraft: gravity/weight and drag (friction of the aircraft acting against the air)

thrust and lift are also forces that operate on an aircraft in flight

1.2.1

(1) payload

(2) consumes

(3) adjust

(4) overcome

(5) opposing

(6) denotes

(7) resists

(8) magnitude

(9) overcome

(10) generate

(11) varied

1.2.2

1) Weight is a vector quantity (with directional element) and mass a scalar quantity (magnitude only). A person's weight is different on the Moon than on Earth, for example, due to the difference in gravity. That same person's mass would be the same on both planetary bodies.

2) Speed is a scalar quantity and velocity a vector quantity, although in everyday English this distinction is not maintained.

3) Perpendicular refers to two objects that meet at a right angle to each other, e.g. the walls of the room are perpendicular to the floor. It is important to note and remember the full verbal expression: "A **is perpendicular to** B"

1.2.3

Many translations are possible. The following is one such translation.

Four forces always act on an airplane in flight: lift, gravity (the weight of the plane), thrust and drag. As can be imagined, lift and gravity oppose one another and the same applies to thrust and drag. Lift is the most essential element when flying. During flight, other forces affect the system and their interplay results in the flight behaviour of the aircraft.

1.3.1

Important vocabulary:

Anstellwinkel = angle of attack

Einstellwinkel = rigging angle

1) Lift is the upward force that keeps an airplane aloft.

   Drag is the frictional resistance offered by the air as the aircraft moves forward.

2) Lift is produced by the difference in pressure created by the airflow moving faster along the upper part of an airfoil in order to meet the airflow travelling along the bottom of an airfoil at the trailing edge at the same time.

3) Displacing the centre of pressure by changing the angle of attack affects the stability of an aircraft.

4) The wing shape of a cargo plane is thicker than that of a faster aircraft in order to provide more lift. If the airflow along the top of the airfoil has further to travel, the airfoil accelerates thereby lowering the pressure along the top of the airfoil and increasing the lift.

5) The function of a propeller is to convert the torque developed by the engine into thrust to propel the aircraft forward.

6) With a variable-pitch propeller the pitch setting can be changed while the propeller is rotating, which allows the efficiency of the propeller to be maintained while the forward velocity changes.

1.3.2

a) aloft = in the air

b) loft is related to the German word „Luft“

c) ashore, abroad, asleep, afloat, alive

d) decelerate (Notice that there is only one /c/.)

e) aircraft: The plural form is invariable, i.e. it does not change. The word aircrafts with an / s / does not exist.

f) see notes under 12.4

g) examples: “The price of commercial space flight is so high as to deter most companies from pursuing this research.”

   “The delamination of composite materials can be so severe as to prevent an airplane from taking off.”

h) maintain = keep The word “maintain” is also a synonym for “service, repair” as in “The aircraft are maintained by the mechanics on a regular basis.”

i) torque = the force of an object rotating about an axis

j) to propel (verb); propulsive (adjective), propelling (adjective)

1.3.3

This course deals with the fundamental **concepts** in aerodynamics. Students learn how **airfoils** produce lift, and how the pressure **distribution** about an airfoil can be **calculated**. The course also introduces the boundary-layer concept, how boundary **layers** lead **to** drag, and what makes them prone to instability or **separation**. It covers the effects of the wing planform shape on lift and drag and includes an introduction to airfoil design and high-speed **aerodynamics**.

1.4.2

There are several theories that explain how airplanes fly. Some are more controversial like the "Skipping Stone Theory". Generally, there is agreement that using the Bernoulli Principle and Newton's Third Law of Motion we can explain how lift is generated.

The Navier-Stokes equations are the fundamental equations of fluid mechanics, the branch of science dealing with the movement of fluids. As the air around us is a fluid, it is useful to consider fluid mechanics in the explanation of how lift is generated. The Navier-Stokes equations have many applications, including the design of aircraft.

1.4.3

1) The wings produce lift if the aircraft moves **through** the air.
2) Engines are most often directed so that thrust works in **the** opposite direction **to** the velocity of flight.
3) The air on the upper side moves faster. **Thus** the pressure becomes smaller.
4) The movement of the wing, and **thus** the plane, has to be fast enough to produce a difference in pressure.
5) This airfoil should produce **as little drag as possible**.
6) Lift is an upward force that overcomes gravity. (no definite article before "gravity")
7) Drag and thrust **oppose** each other.
8) The difference in pressure generates the force mentioned **in/at** the beginning.
9) But this **principle** is only possible when the aircraft is moving forward.
10) **With** conventional aircraft designs, the ailerons and flaps are also attached to the wings.
11) The wings of an airplane produce the **main** part of the lift.
12) Engines on an airplane **produce** the necessary thrust to overcome drag.
13) Wings accelerate the air flow **along/across** the upper part of the airfoil. ("On" is also possible, but along/across are preferable.)
14) Drag must be opposed by propulsive thrust, as the plane otherwise would decelerate and **lose** height (The word "altitude" is preferable to "height" here.)
15) On the upper side of the wing, the air has to **travel** a larger distance.

16) The air moves faster and this results **in** an area of low pressure.

17) Lift results **from** the difference between the pressure at the upper and lower surface of an airfoil.

18) Because of the higher velocity on the upper surface **there is** lower pressure.

19) You must reach a velocity which **makes it possible to lift the aircraft and to keep it** aloft.

20) **The function of engines is to make** the plane move forward.

1.5.1b

b) James Bond double oh seven

treble seven five six seven two OR triple seven five six seven two

1.5.6

a) one half

b) two thirds

c) three quarters

d) four fifths (don't forget to pronounce the /s/)

e) seven ninths

Chapter Two: Wings and Airfoils

2.1.1

A wing planform

B leading edge

C trailing edge

D wing area

E span

F chord

G centreline

H aspect ratio

I airfoil

J chord line

K thickness

L camber

M mean camber line

N tip

O dihedral angle

Aspect ratio indicates the relationship between the length of an airfoil from root to tip compared to its thickness. The higher the aspect ratio the longer and thinner the airfoil. A high aspect ratio means that there is less drag acting on the airfoil. Gliders have, for example, high aspect ratios. A low aspect ratio means that the span of the airfoil is relatively short or that the airfoil is wide (i.e. has a long chord). Therefore, one can say that there is an inverse relationship between aspect ratio and drag.

2.1.2

A slat

B main airfoil

C vane

D flap

E slots

2.1.3

Advantages:

- airplane can fly at higher speeds, i.e. supersonic, whereas a straight wing aircraft cannot fly much beyond the speed of sound
- less induced drag

Disadvantages:

- lower lift than a straight wing aircraft
- spanwise flow at the tips of the airfoil (some of the air moves sideways along the wing from the root to the tip because of the angled wing)

2.2.1

1) Mini-TEDs are multifunctional trailing edge devices that are 9 to 15 centimetres wide and made of carbon fibre composites. They are integrated into the underside of the landing flap.

2) Benefits include greater lift, which enables quicker take-offs and steeper angles of approach when landing. The noise footprint is, therefore, lower. Moreover, the wing can be made lighter because lift distribution can be better controlled during turbulence. At high airspeeds, drag can be lowered by redistributing lift using mini-TEDs. This results in less fuel consumed and lower emissions.

3) The airflow is directed downwards and produces greater lift. Mini-TEDs increase the critical angle of attack at which stall occurs.

4) take-off, landing, high speed cruising, turbulence, etc.

2.2.2

highly cambered profile = thick airfoil

retractable slats and flaps = additional elements at the leading and trailing edges of the airfoil that can be moved in and out

spoilers = hinged section in the middle of the upper-side of an airfoil that is used to reduce lift and increase drag when landing, for example

test-rig = set of equipment used for testing

slipstream = a vortex of air at the trailing edge of an airfoil

stall = airflow becomes separated from airfoil causing a loss of lift due to the maximum angle of attack being surpassed

noise footprint = amount of noise produced; modelled on the expression "environmental footprint"

gusts = sudden, strong winds

pollutant emission = gases that are produced during combustion and exit from the jet engines

2.2.3

exhibits = shows

detracts = takes away from something otherwise positive

employed = used

sound = reasonable

to accommodate = to take account of; here: to adjust to

auxiliary = support, additional, back-up

decisive = crucial, essential, important

successor = next, later, follow-up

scope = area

to subject sth. to sth. = to expose sth. to sth.; to cause sth. to experience sth.

alleviate = lessen, lower

deployment = operation, use

2.2.4

possible features could include: strut-braced wings, hydrogen-powered aircraft, blended body wing, perforations in the airfoil to provide for smooth laminar flow over the airfoil

2.3.1

x plus y equals z

x minus (subtract/take away) y equals z

a multiplied by (times) b equals c

a divided by b equals c

addition, subtraction, multiplication, division

2.3.2

a squared plus b squared equals c squared

a cubed

a to the power minus n (or a to the power of minus n)

a to the power n minus two

n times x sub one minus m times x sub two

2.3.3

the square root of a plus the cube root of b equals c

the fourth root of a

the nth root

2.3.4

open brackets x plus y close brackets squared times open brackets a minus c close brackets cubed equals z

OR

x plus y in brackets squared times a minus c in brackets cubed equals z

b equals open square brackets open brackets a plus c close brackets minus open brackets d plus e close brackets close square brackets

the set two, four, six, eight (Here, the brackets are called braces, but we do not say the word "braces".)

Symbol	Reading	German Translation
$\pm$	plus minus (plus or minus)	plus/minus
$>$	greater than	größer als
$<$	less than	kleiner als
$\geq$	greater than or equal to	größer gleich
$\leq$	less than or equal to	kleiner gleich
$\approx$	approximately equal to	ungefähr gleich
$\neq$	is not equal to	ist ungleich
$\rightarrow$	approaches	strebt gegen
$\therefore$	therefore	folglich (represented as "$\Rightarrow$" in German)
∞	infinity	unendlich

2.4.2

1) *x* to the power of minus *p* equals one over *x* to the power *p*
2) *x* equals *n* times *x* sub one plus *m* times *x* sub two all divided by m plus *n*
3) *y* does not equal ten thousand and one
4) *b* squared is greater than or equal to a squared times open brackets one minus *e* squared close brackets.
5) the cube root of *x* plus the *n*th root of six equals *y*

Chapter Three: Aircraft Motion

3.1 Pitch Motion

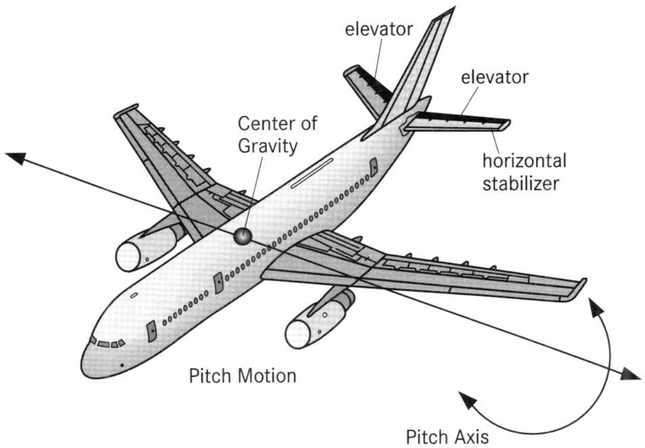

Fig. AK 1

3.2 Roll Motion

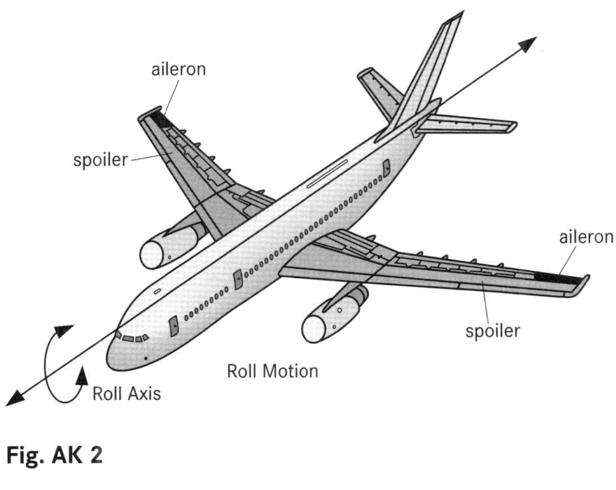

Fig. AK 2

3.3 Yaw Motion

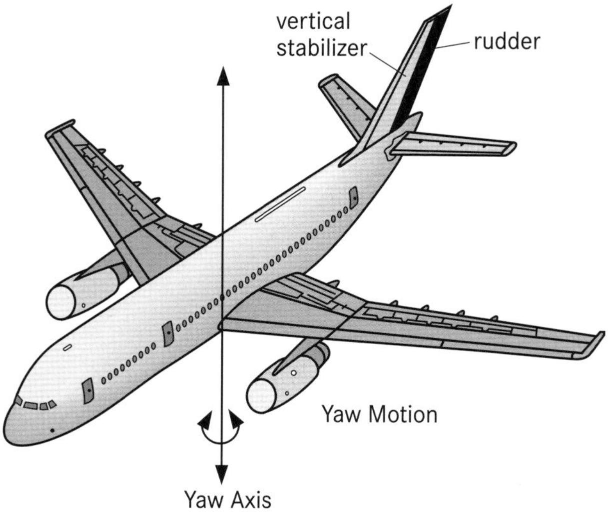

Fig. AK 3

3.4.1

1) **Trimming** involves adjusting the movement and position of the aircraft in order to achieve a stable flight in which there is no movement around the centre of gravity.
2) **Trim systems** relieve the pilot of having to keep constant pressure on the flight controls.
4) An **antiservo tab** keeps the stabilator in the desired position. Adjustable tabs prevent an airplane from skidding left and right during normal flight.

3.4.2

crank = Trimkurbel

hinged = mit einem Scharnier verbunden

rod = stange (here: Steuerflächenstange)

skid = schleudern, rutschen

pivot = sich drehen

rear spar = Hinterholm

jackscrew = Spindel, Nivellierschraube

3.5

1) the absolute value of x
2) n factorial OR factorial n

3) the limit of *f* of *x* as *x* approaches infinity

4) the definite integral of *f* from *b* to *a*

5) the sum of *as d* runs from 1 to *n* of *B* sub *d*

6) the null set

7) *a* approaches *b*

8) the first derivative with respect to *x*

9) the *xs* cancel

10) the intersection of X and Y

Chapter Four: Propulsion Systems

4.1.1

1) Verbrennungsmotor, Kolben, Kurbelwelle

2) the air

3) to compensate for the fact that the tips move faster than the hub as the blades rotate

4) A turboprop is an engine that turns a propeller.

4.2.1

1) Both engines use a hot exhaust gas that is passed through a nozzle to produce thrust.

   A rocket engine carries its own oxygen for combustion, while a turbine gets its oxygen from the surrounding air.

2) Düse, Abgase, Hochgeschwindigkeitsflug

3) A Boeing 777 can travel at high speeds, whereas an Airbus A400M, a cargo aircraft, does not have this requirement.

4.3.1

1) There are no compressors in a ramjet. Therefore, they are lighter and simpler than turbojets. Ramjets can only produce thrust when the vehicle is already moving.

2) Ramjet comes from the verb "to ram", which means to force something with a lot of force into a certain space. In a ramjet, the external air is rammed into the combustion chamber using the forward speed of the vehicle.

3) **subsonic** = below the speed of sound = unter Schallgeschwindigkeit, Unterschall ...

   supersonic = above the speed of sound = supersonisch, Überschall ...

   transsonic = around the speed of sound = schallnah

4.4.1

1) Whereas turbine engines use the surrounding air as their working fluid, rockets use combustion exhaust gases. In outer space there is no atmosphere, so turbine engines cannot work there.

2) **sound barrier** = Schallmauer

   exceeded = überschritten (to exceed = überschreiten)

   propellant = Treibstoff

3) Liquid rockets tend to be more complex because of the pumps and storage tanks, which allow the flow of the propellant to be regulated even after the rocket has been fired.

4.7.1

Number of Shape	Noun	Adjective
1	square	square
2	triangle	triangular
3	circle	circular
4	rectangle	rectangular
5	ellipse	elliptical
6	cylinder	cylindrical
7	cube	cubic
8	rectangular prism	rectangular
9	octagon	octagonal
10	pentagon	pentagonal
11	cone	conical
12	sphere	spherical
13	semi-circle/semicircle	semi-circular/semicircular
14	hemisphere	hemispherical

4.8.1

1) acute angle
2) chord
3) perpendicular to
4) length
5) width
6) external angle
7) breadth
8) height
9) parallel
10) circumference
11) bisector
12) radius

13) obtuse angle
14) diameter
15) perimeter
16) right angle
17) arc
18) area

Chapter Five: Composite Materials

5.1

1) malleable
2) conductive, ductile
3) reactive
4) semi-conductive
5) transparent, smooth
6) brittle
7) inert, noble
8) composites
9) dull

5.2.2

1) Composites are composed of a polymer resin matrix that contains layers of strong reinforcing carbon or glass fibres.
2) They are being used because they are lighter, which means less fuel consumed and fewer pollutant emissions. They are also stronger and can withstand greater stresses than conventional materials.
3) They fatigue and fail in a different manner from metals. When a crack occurs, it will reach a critical size much more quickly than with metals, such as aluminium.
4) a) tap test = tapping sections of the fuselage and listening for any changes in the sound that would identify a crack

   b) thermal cameras and ultrasound to identify any liquids between the composite layers

   c) lasers that would heat the fuselage section and sensors that would identify any changes in frequency of the vibrations

   Visual inspections are unsatisfactory because the cracks or delaminations occur below the surface and are therefore not visible to the naked eye.
5) tail fin, rudder, wing box, fuselage hull, wing sections
6) Think about A350 or Boeing 787.

5.3

Atomic Number	Symbol	German	English
1	H	Wasserstoff	hydrogen
6	C	Kohlenstoff, Karbon	carbon
7	N	Stickstoff	nitrogen
8	O	Sauerstoff	oxygen
9	F	Fluor	fluorine
11	Na	Natrium	sodium
13	Al	Aluminium	aluminium (British and international English) aluminum (American English)
14	Si	Silizium	silicon
16	S	Schwefel	sulfur (American English) sulphur (British English)
17	Cl	Chlor	chlorine
19	K	Kalium	potassium
22	Ti	Titan	titanium
26	Fe	Eisen	iron
28	Ni	Nickel	nickel
29	Cu	Kupfer	copper
30	Zn	Zink	zinc
32	Ge	Germanium	germanium
50	Sn	Zinn	tin
74	W	Wolfram	tungsten
80	Hg	Quecksilber	mercury
82	Pb	Blei	lead
92	U	Uran	uranium

1) **Atomic number** indicates the number of protons that an element has.
2) There are three main subatomic particles: **electrons**, **protons** and **neutrons**. Protons and neutrons are found in the nucleus of an atom, whereas electrons orbit around the nucleus. Protons and neutrons can be broken down into quarks.

Chapter 6: Helicopters

6.1.1

1) A helicopter takes off when the angle of attack of its rotating blades reaches a certain value that produce enough lift to overcome the weight of the vehicle.
2) Horizontal flight is achieved by tilting the rotor forward at a certain angle using cyclic pitch change, which involves changing the pitch of each blade once per revolution.
3) Swash plates swivel the blades back and forth as they rotate thereby enabling the pitch to be changed cyclically.
4) Blade flapping is the upward and downward movement of the blades, which eliminates the stress on the blades caused by the tilting of the rotor while the blades rotate.
5) A single-rotor helicopter needs a small propeller at its tail to prevent the helicopter from rotating in the opposite direction to rotation of the blades.

6.1.2

a) instead of

b) attain = reach; obtain = receive, get

c) The helicopter blades rotate quickly thereby generating enough lift to allow the helicopter to lift off the ground.

d) plane = flat surface

e) hinges = Scharniere

f) regardless of, independent of

6.1.3

1) revolution, rotation, swivel, roll
2) revolution, roll
3) wobbling, to and fro, flap
4) propel, to and fro
5) propel, revolution, flap

6.1.4

Several translations are possible. The following translation is one such possibility.

In order to switch from hovering to a vertical climb, the rigging angle of all the rotor blades must be increased evenly by using the collective pitch. An increase in the rotor thrust and the lift brings about an increase in the blade resistance, which has to be overcome by greater engine power. The adjustment of the engine power can be carried out by a twist grip throttle as well as automatically through engine power regulators. As soon as the helicopter has switched to an even vertical climb, the rotor thrust can be reduced to a level that corresponds to the weight and the resistance of the cabin.

6.2.1

1)

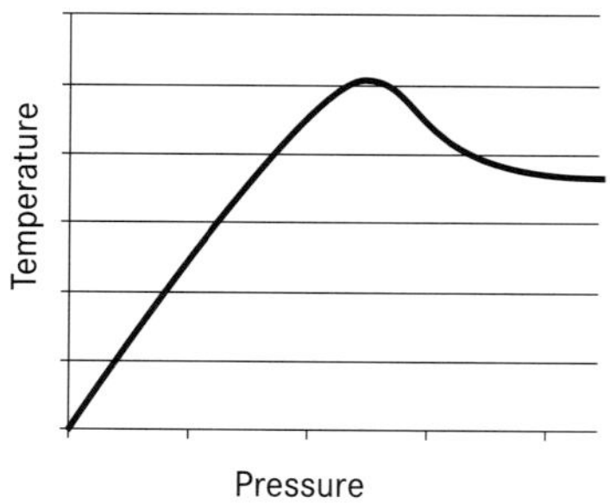

Fig. AK 4

3)

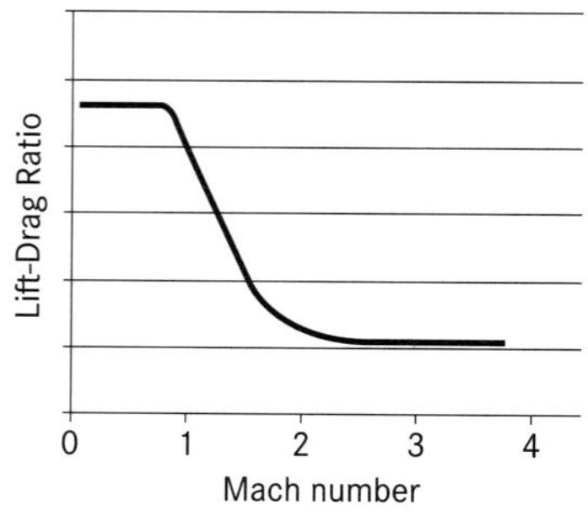

Fig. AK 5

2)

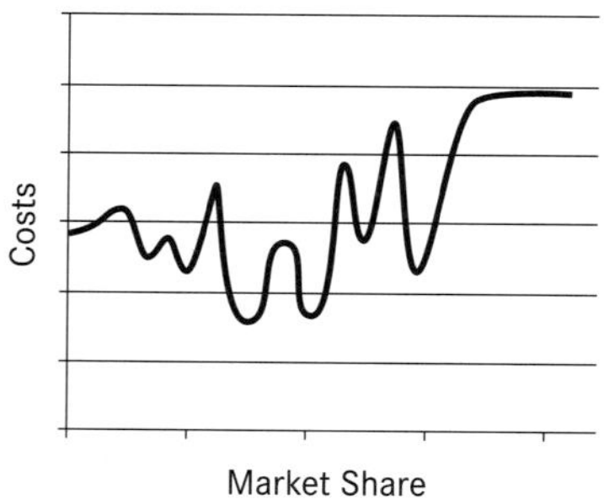

Fig. AK 6

6.3

Expressing agreement
I couldn't agree more.
I think so, too.
I'm of the same opinion.
Saying you partly agree
I agree (with you) up to a point, but...
To a certain extent, I agree with you. However, ...
Disagreeing in a less direct way
I'm not sure I agree with you.
I beg to differ with you.
Disagreeing in a more direct way
That argument just doesn't hold.
I don't agree with that at all.
Saying you are undecided
I'm of two minds about that.
I'm not sure where I stand on that.

Chapter 7: The Future of NASA

7.3.1

a) 8

b) 9

c) 2

d) 4

e) 13

f) 10

g) 3

h) 5

i) 6

j) 11

k) 12

l) 1

m) 7

Note: bellows = Kompensator, Längenausgleich

7.3.2

1) because of repairs; huge orange fuel tank still needs work to reduce chances that hazardous debris will fall from it during lift-off

2) a chunk of foam fell off the external tank during lift-off and punched a lethal hole in the shuttle's wing

3) ice

4) bigger risk than thought; because of ice five-inches long and travelling at three times the speed of sound

   ice develops on humid days in Florida when cold liquid oxygen hits bare metal and condenses

5) developed bellows heater to prevent ice formation on bare metal; bellows heater used in addition to foam

6) to finish assembling the space station

7) to repair the ailing Hubble telescope

7.4.2

Criterion	Space Shuttle	Orion Spaceship
Purpose / Function	- developed to build the ISS and launch space telescopes (e.g. Hubble) - can bring payload and crew to low-earth orbit and back	- being developed for human space exploration and exploration of asteroids, the Moon and eventually Mars - will be able to launch payload and crew into space, but only crew will return in the capsule (cannot retrieve payload and bring it back to Earth)

Criterion	Space Shuttle	Orion Spaceship
Design / Features	- looks like an airplane - can land on a runway - shuttle is attached to external rocket tank, which raises the risk of serious damage to the shuttle due to falling foam and ice - reusable, but expensive on account of new boosters for each launch - one rocket needed - no safety/abort mechanism during lift-off - old, but proven technology	- perched on top of an Ares I rocket (no risk from falling debris or ice) - capable of landing on the ground (instead of the water) by using parachutes, steering rockets and airbags - can be reused ten times - two rockets needed: Ares I for the crew and capsule, Ares V for payload - safety/abort mechanism has been designed - state-of-the-art technology
Capacity	- seven crew members to low-earth orbit (LEO) - greater cargo capacity	- six crew members to LEO and four to the Moon - lower cargo capacity
Costs	- maintenance of the shuttle fleet has been very expensive	- costs expected to be lower
Additional Features		- outfitted with real toilets!

7.5

a) **isolated** = isoliert; isoliert can also be translated as insulated as in insulated cable

The pipes are insulated to prevent heat loss.

b) **construction** = Bau...; Konstruktion = design

The construction of the new building was delayed on account of bad weather.

c) **control** = steuern, beherrschen; kontrollieren = inspect, check

Mini-TEDs are some of the many parts used to control the aircraft during take-off and landing.

d) and e) **technique** = Ausführungsart; Technik = technology, engineering *or* technique

The company has invested more money in technology to improve productivity in the production department.

Students of aerospace and mechanical engineering generally earn quite well after graduation.

Scientists are developing a new technique to synthesize human cells.

f) **beamer** = does not exist in English (in American slang it stands for a BMW); Beamer = (digital) projector

g **eventually** = schließlich; eventuell = maybe, possibly

Eventually composite materials will be used more and more in aircraft manufacturing.

h) **actually** = tatsächlich, eigentlich; aktuell = current, recent, up-to-date

Greece's deficit was reported to be 10 per cent of GDP. Actually, the deficit is far higher.

i) **special** = besonders; speziell = particular, specific

The Dreamliner is a special aircraft, given that its various parts were manufactured at plants around the world and then assembled in the US.

j) **sensible** = vernünftig; sensibel = sensitive

I would agree with your proposal; it seems sensible to me.

k) **prove** = beweisen; prüfen = check, test

How and why airplanes fly hasn't been fully proven up to now.

l) **promotion** = Beförderung *or* Reklame, Werbung; Promotion = dissertation, PhD

Our latest engineer hired received a promotion only eight months after starting his job, as he was so good at his work.

m) **consequently** = infolgedessen, somit; consequent = consistently

Using composite materials will lower fuel consumption. Consequently, airlines' costs will be lowered as well.

n) **prospects** = Aussichten; Prospekte = brochures

The job prospects for university graduates have improved since the latest recession.

o) **defect** = Makel; defekt = broken, defective, damaged

There was a defect in the diamond, which, of course, lowered its value.

p) **delicate** = feinfühlig, zart *or* heikel; delikat = delicious

He is rather delicate, so I would take great care when giving him bad news.

q) **floor** = Etage or Boden; Flur = hallway, corridor

Our hardwood floors give our office a warm atmosphere.

r) **edge** = Kante; Ecke = corner

A knife's edge is usually sharp.

Chapter 8: Space Elevators

8.1.1

1 g, 2 f, 3 p, 4 c, 5 b, 6 i, 7 m, 8 h, 9 e, 10 n, 11 j, 12 o, 13 d, 14 k, 15 a, 16 l

8.1.2

1) A **space elevator** is a cable extending from a satellite orbiting in space down to Earth used to transport payload into space.

2) A climber would ascend the cable and release the payload into orbit around the Earth.

3) The costs of launching payloads into space would be much lower than with conventional means. It would cost only $100/kg instead of $20,000/kg.

4) The tension on the cable has been one of the most serious difficulties with this design concept. The cable would need to support 100 GPa.

5) Buckyballs or buckytubes, which are made of carbon nanotubes, are currently being examined as a possible material for the cable.

6)

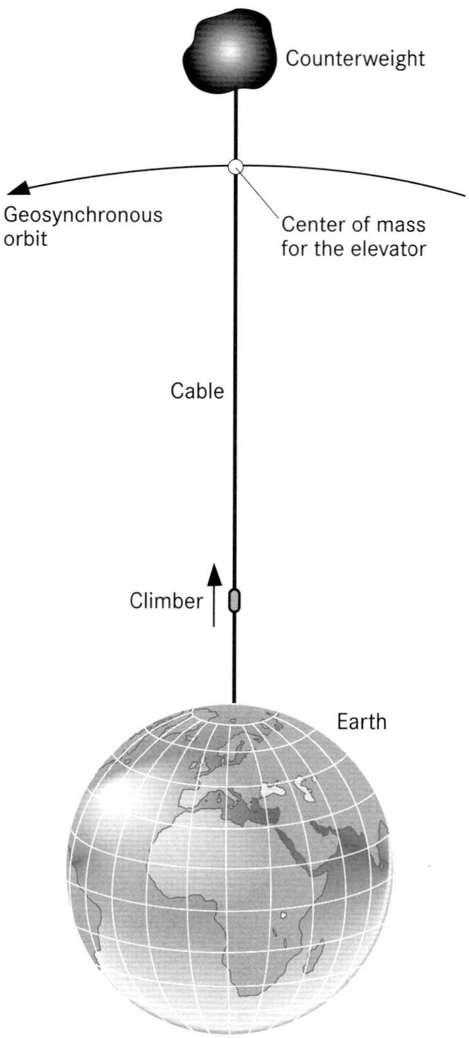

Fig. AK 7

8.4

uncountable nouns: furniture, time, money, coffee, mail, information, advice, news, luggage, happiness, love, etc.

Chapter 9: Electrical Rocket Propulsion

9.1.1

fundamentally	grundlegend
performance	Leistung
propellant, fuel	Treibstoff
thence (rare) = therefore	deshalb
essentially	im Wesentlichen
inertial	träge
charged	geladen (elektrisch)
conductive	stromführend
thruster	Schubdüse
fission reactor	Kernspaltungsreaktor
nozzle	Düse
incidental	nebensächlich
azimuthal	umfangsgerichtet
constrict	behindert; beschränkt
circumvent a problem	ein Problem umgehen

9.1.2

1) **Benefits**:

   - not energy-limited; they can reach higher energies per propellant used
   - possible to achieve high exhaust velocities
   - higher specific impulse than chemical or nuclear rockets (i.e. more efficient)

   Drawbacks:

   - limited by the mass of conversion equipment
   - high power consumption
   - low thrust

2) because gravitational forces are lower and are nearly balanced by inertial forces; longer journey times

3)

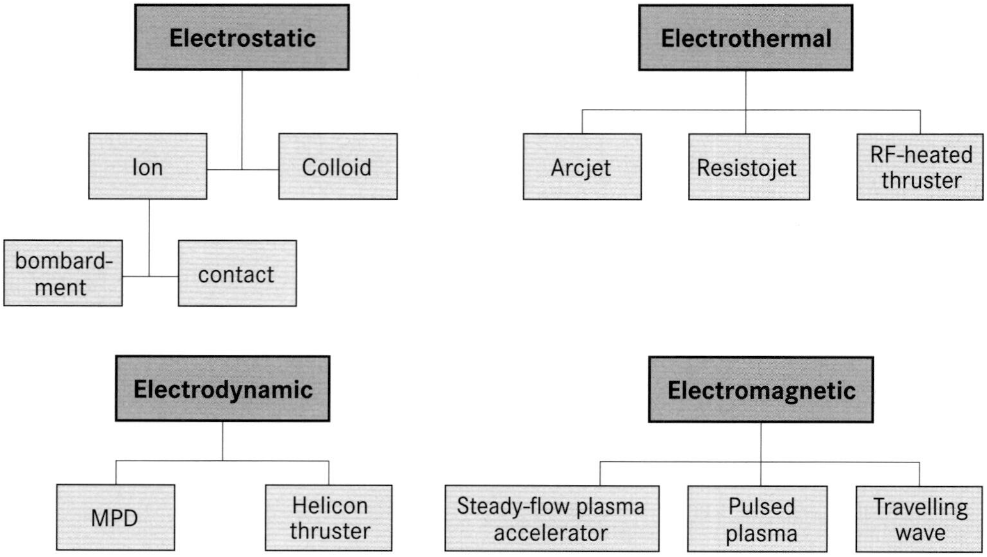

Fig. AK 8

9.2.1

a) 2
b) 4
c) 7
d) 10
e) 1
f) 8
g) 9
h) 5
i) 3
j) 6

9.2.2

1) They can be considered electrostatic because the ions are accelerated in the propellant. They are also electromagnetic because the plasma electrons interact with the magnetic field to create an electromagnetic field

2) SPTs

- length and diameter of chamber are similar
- dielectric (insulator than can sustain a steady electric field) and ceramic material
- TALs
- very short acceleration channel
- no insulating material

3)

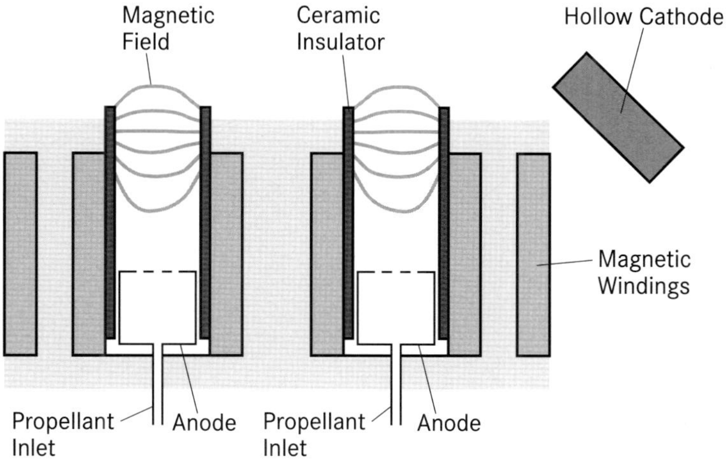

Fig. AK 9

9.3

Although the thrust of the electrical propulsion system on SMART-1 was considerably lower than that of a small monopropellant thruster or a Fregat main engine, the thrust time was considerably longer, thereby allowing SMART-1 to reach significantly higher velocities over time. Furthermore, the efficiency of the propulsion system, as expressed by specific impulse was far greater than that of the other two propellants (see following equation for the specific impulse expressed in seconds):

$I_{sp} = C_e/g_0$

C_e = exit velocity of the propellant (mass/seconds)
g_0 = gravity on the ground (mass/seconds^2)

Lastly, the amount of propellant consumed by the electrical propulsion system in SMART-1 was comparable to that of a small monopropellant thruster, but markedly lower than the amount of propellant consumed by the Fregat main engine. This would lower the weight of the space probe and thereby the costs.

9.4

And

also

in addition

furthermore

additionally

moreover

not only ... but also

as well as

But / In contrast

on the other hand

though

whereas

instead

however

although

on the contrary

in comparison

unlike

as opposed to

despite

yet

For example

such as

for instance

Because

in order to

due to

because of

that is the reason why

since

As a result

resulting in

thus

therefore

consequently

hence

In conclusion

to summarise

in summary

Chapter 10: Orbits and SMART-1

10.1

1) **S**mall **M**issions for **A**dvanced **R**esearch in **T**echnology
2) on the night of Sept. 27-28, 2003; it was launched atop an Ariane 5 rocket; it was one of three payloads
3) to test solar electric propulsion for long-distance space missions; to test the miniaturisation of spacecraft and instruments (only 1 cubic metre)
4) infrared rays to map lunar minerals and search for water; X-rays to gain information about chemical elements on the Moon

5 condensed Earth debris vapourised during asteroid collision millions of years ago; gravity solidified and condensed the debris/dust over the years

10.2.1

1) **Lunar resonances** are when the period of the Moon is aligned with that of the space probe. The space probe encounters the gravitational influence of the Moon, which is used to change the orbit or trajectory of the space probe.
2) **L1** denotes the point where the gravitational effects of the Moon and the Earth are in balance. This point is 50,000 to 60,000 km out from the Moon in the direction of the Earth.
3) During a planetary fly-by the trajectory of the spacecraft acquires a major change from gravitational interactions with the larger body.
4) A **semi-major** axis is one half of the longest distance of an ellipse, whereas a **semi-minor** axis is one half of the shortest dimension of an ellipse.
5) A geostationary transfer orbit is an elliptical orbit between two circular orbits. Specifically, it is a Hohmann transfer orbit around the Earth between a low Earth orbit and a geosynchronous orbit.

10.2.2

circles; sidereal; sunset; accounts; technically; direction; sidereal; adjusted; equator; inclination; pole; travels

10.2.5

ihre Geschwindigkeit wird nachlassen ...

der anhaltende Schub des elektrischen Raumfahrtantriebs ...

die Anziehung der Schwerkraft des Mondes ...

die Wirkung der Schwerkraft war viel ausgeprägter ...

10.3

1) Xenon is the propellant and solar energy from the solar arrays provides the electricity.
2) ESA wanted to test this key new technology. They wanted to test the suitability of this type of propulsion system for longer trips, as chemical rockets would not be possible to travel to Mercury. In particular, they wanted to test how the system interacts with electric and magnetic fields in space.
3) Given its steady acceleration of 3 to 4 km/s, an electric propulsion system can ultimately achieve much higher velocities over time, as it is not energy limited. With a conventional propulsion system, a maximum velocity is reached over a certain amount of time and the spacecraft ceases to accelerate.

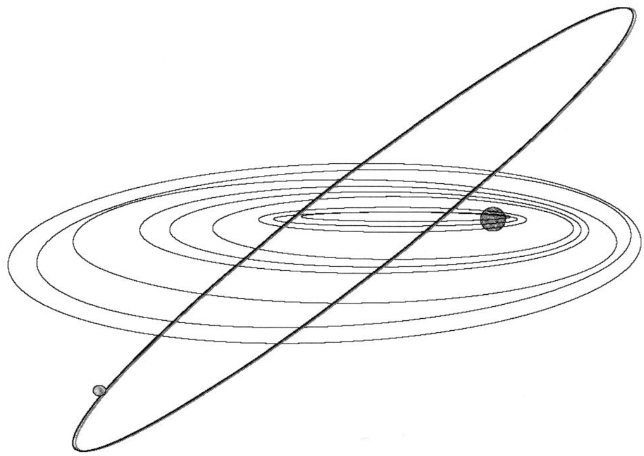

Fig. AK 10

10.5.1

Cause	Effect
to cause	to be affected by
to result in	to be the consequence of
to affect	to result from
to lead to	to be due to
to be the cause of	to be the result of
to bring about	to be caused by
to produce	to be produced by
to influence	to be brought about by

10.5.2

Several answers are possible.

a) The increased velocity of the airflow along the top of an airfoil produces/generates lift.

b) Higher fuel consumption is the result of greater drag.

c) The angle of attack influences lift.

d) Fuel combustion causes thrust.

e) The pitching motion of an aircraft is brought about by the deflection of an aircraft's elevator.

f) The weakening of composite materials results from ageing and stress.

g) A space elevator would lead to lower launch costs.

Chapter 11: Commercial Space Flight

11.2

1) two trips to space and back in 14 days; must reach 62 miles above the Earth and be privately funded

2) very simple; looks like a 45-calibre bullet; 2 m in diameter and 8 m long; matt black exterior; only two tiny windows; doesn't look like a rocket

3) hybrid rocket engine (liquid and solid rocket fuels); a balloon is floated to 80,000 feet and the craft is tethered to the balloon 750 feet below it; the craft is then launched from this altitude

4) Wild Fire

5) less than $300,000

6) not launched from specially designed aircraft; smaller budget; volunteer effort; simplicity and innovation

7) develop a larger craft that will carry eight passengers into space; would not go into orbit, but rather high enough to experience weightlessness and witness incomparable view of the Earth

11.3

In principle, SpaceShipOne is not designed for orbital flights, as its maximum speed amounts to only approximately one seventh of the necessary satellite velocity.

The plane will reach a speed of Mach 3.5 during the 65 seconds the engine is firing. It has a hybrid rocket engine built by the American company SpaceDev, whose fuel/propellant consists of a solid fuel (HTPB) – a type of tire rubber – and a liquid oxygen fuel. It is partly reusable. The ballistic space flight lasts about three and a half minutes. Then, the upper layer of the atmosphere is reached again. During take-off, the plane is towed by a carrier, the so-called White Knight, to an altitude of about 15 km where it is released. After that the rocket engine is fired.

1. kosmische Geschwindigkeit = satellite velocity

2. kosmische Geschwindigkeit = escape velocity

HTPB = hydroxyl-terminated polybutadiene

Chapter 12: International Space Station

12.1.2

truss segment = part of a zigzag-shaped or triangular framework (Teil eines Kragträgers)

cupola = dome (Kuppel)

mating adapters = a part that allows spacecraft of different sizes to dock with each other (Kopplungsadapter)

logistics carrier = a small spacecraft used to transport equipment (Logistikträger)

attached exposed exterior platform = a platform that is outside but attached to the ISS

vehicle-sustaining engineering = technology and know-how to keep the ISS running over time (sustaining = nachhaltend)

NASA launched the satellite **into** *orbit.*

The satellite is **in** *orbit around the Earth.*

port = left side of a ship or spacecraft (Backbord)

starboard = right side of a ship or spacecraft (Steuerbord)

solar array = a group of solar panels connected together

inclination = angle between the orbit of the ISS and the equator (Bahnneigung)

The semi-colons are used to list several items in which there are already commas. Normally commas are used to list items, but in order to prevent confusion and to clearly delineate where one item starts and where it ends, semi-colons are used when the items themselves contain commas.

For example:

"The ISS consists of several components: a Japanese laboratory, which has an exterior platform; a spacecraft for crew return, the Soyuz; living quarters..."

Here we can see that there are three items separated by semi-colons because the first and second items contain commas.

12.3.3

maintaining conditions

performing maintenance

Student's Lexicon

As you go along, be sure to record not only the new words you find in the texts provided, but also the groups of words or „collocations" and the context or sentence in which you found them. This will help you remember the word as well as use it in the correct situation.

A few words have been provided for you. Use the remaining space to supplement this list as you go along.

Word	Meaning / Translation	Example Sentence / Collocation
Chapter One		
aloft	in the air	
airfoil	similar to wing	
drag	Widerstand	
to be prone to	anfällig sein für	

Word	Meaning / Translation	Example Sentence / Collocation
Chapter Two		
dihedral angle	V-Stellung	
detracts from	takes away from	"This detracts from the optimum performance."
Chapter Three		
yaw motion	Gierbewegung	

Word	Meaning / Translation	Example Sentence / Collocation
Chapter Four		
crankshaft	Kurbelwelle	
perpendicular	senkrecht / orthogonal	"The horizontal stabilizer is **perpendicular to** the vertical stabilizer."
Chapter Five		
translucent	lichtdurchlässig	
pitfall	disadvantage / danger	

Word	Meaning / Translation	Example Sentence / Collocation
Chapter Six		
swash plate	Taumelscheibe	
in lieu of	instead of	
Chapter Seven		
debris	Trümmerschutt / Bruchstücke	
insulated	isoliert	"insulated wires"

Word	Meaning / Translation	Example Sentence / Collocation
Chapter Eight		
buckyballs / bucky-tubes	Fullerene	
Chapter Nine		
to a large extent	weitgehend	
thus		"Electrical propulsion systems have a high specific impulse. Thus, they are better suited to long space missions."

Word	Meaning / Translation	Example Sentence / Collocation
Chapter Ten		
sidereal day	a day measured in relation to a distant star	
technically speaking	strenggenommen	
Chapter Eleven		
feasible	workable / doable	"Which aircraft design is the most feasible?"

Word	Meaning / Translation	Example Sentence / Collocation
Chapter Twelve		
starboard	Steuerbord	